Stay Healthy by Supplying What's Lacking in Your Diet

David Coory

ZEALAND PUBLISHING HOUSE
Private Bag, Tauranga. New Zealand.

© 2002 ZEALAND PUBLISHING HOUSE

Sections of up to ten pages of this book may be reprinted for non-profit purposes without permission of the publishers provided acknowledgment of the source is clearly stated.

First published	1988	by Zealand Publishing House as *"NZ Nutrition and Your Health."*
2nd Edition	1989	Reprint.
3rd Edition	1990	Revised and updated.
4th Edition	1992	Revised and updated. Retitled *"Stay Healthy by Supplying What's Lacking in Your Diet."*
5th Edition	1995	Reprint.
6th Edition	1996	Revised and updated.
7th Edition	Jan 2001	Revised and updated. New cover.
8th Edition	Sept 2001	Revised and updated. Reprinted Feb 2002, August 2002.

Further copies of this book can be obtained by writing to:
Zealand Publishing House
Private Bag 12029,
Tauranga. New Zealand.

Or by Phone, Fax, Email, or Internet (credit card required).

Phone   0800 140-141 (NZ only) or (07) 576-5575
Fax   0800 140-141 (NZ only) or (07) 576-5510
International: Phone 0064-7-576-5575
              Fax   0064-7-576-5510
Email   sales@zealandpublishing.co.nz
Internet   www.zealandpublishing.co.nz

ISBN   0-908850-09-3

All information, advice and figures in this book are given in good faith and are not intended to replace competent medical advice.

The researcher welcomes comments or corrections that may improve and enhance future editions. Please write to David Coory, Zealand Publishing House, Private Bag, Tauranga. New Zealand, or email david@zealandpublishing.co.nz.

"Health is the best wealth."

Anonymous

This book is dedicated to researchers worldwide, who painstakingly pursue a better understanding of that most complex and marvelous process – the human digestive system.

Acknowledgments

Special thanks to Raymond Coory for his editing, Caren Glazer for her cartoons, Neil Muir for his photography and Publicity Printing for their cheerful professionalism.

Also to all the following who have assisted with helpful comments and constructive criticism. Dr David Consolvo B.Sc, MD, Alex Coory, Ann Braid, Phyllis Hills, Christina Ianussi, Brian Kenyon, Jim Maskell, Tess Maskell, Neil Muir, Mary Muir, David Ollivier, Mary Ollivier, Donald Ross, Arthur Towgood and Amanda Bryan.

Contents

Find me the facts 6
In defence of doctors 8
Layout of the book 10
Measures & abbreviations ... 13
Alcohol 15
Calories 16
Carbohydrates 21
Cholesterol 23
Fats 30
Fibre 37
Protein 40
Sugars 43
Vitamin A 46
Vitamin B1 – Thiamine 49
Vitamin B2 – Riboflavin 52
Vitamin B3 – Niacin 55
Vitamin B6 58
Folate (Folic Acid) 62
Vitamin B12 65
Lecithin 68
Inositol 69
Pantothenic Acid 71
Vitamin C 74
Vitamin D 78
Vitamin E 81
CoQ10 84

Vitamin H – Biotin 86
Vitamin K 88
Aluminium 89
Boron 90
Calcium 92
Chlorine 96
Chromium 98
Copper 100
Fluoride 103
Iodine 106
Iron 108
Magnesium 111
Manganese 115
Molybdenum 117
Phosphorus 119
Potassium 121
Selenium 124
Sodium (common Salt) 128
Sulphur 131
Zinc 134
Conclusions 138
Becoming permanently slim 151
Health and religious beliefs . 157
Food additive code numbers 159
Index 166

Health Hints

The first law of health 18	Back pain and stress 85
Smoking and cholesterol 29	After antibiotics 87
Garlic a natural blood thinner........ 35	Avoiding headaches..................... 87
Avoiding high blood pressure 35	Minimise bed rest.......................... 95
Lower your stress 35	Rid drinking water of Chlorine....... 97
Avoiding constipation and piles ... 38	Use unrefined sugars.................... 99
Muscle power from water 41	Anxiety without obvious cause 101
Minimise colds and flu 42	Copper bracelets and arthritis......102
Tame that sugar craving 45	Avoiding tooth decay....................105
Change your diet gradually 48	Avoiding Asthma 105
Negative thinking and health 51	Minerals in unrefined sea salt 107
Milk a good source of Riboflavin ... 53	How to absorb more Iron 110
Eggs a rich source of nutrients..... 54	Stress and heart disease............112
Cheerful people live 19% longer 56	Breaking an addiction to
Pregnancy after the Pill................. 60	sleeping pills or tranquillisers ...114
Three rules to minimise stress 61	Aerobic exercise 116
A healthy pregnancy..................... 64	Absorbing grains and nuts 118
Milk can supply B12 needs........... 67	The power of affirmations 120
Three ways to a good sleep 69	Five laws of longevity..................122
Preventing acid stomach 72	Lowering the risk of stroke 123
Work off that stress 72	Good Selenium sources 127
The power of our words................ 73	High sweat loss and salt.............130
An apple a day.............................. 76	90 day Arthritis cure 132
Vitamin C a healer 77	Colloidal Silver a natural antibiotic.137
Banish bad breath 80	How fit are you?...........................150
Brisk walking lowers cancer risk 80	Beat the boredom of exercise 156
Anti-oxidant supplements 82	Fasting and health 157

Miscellaneous

Depressing statistic 18
Blood pressure guidelines 28
Cholesterol guidelines 28
Anxiety induced high blood pressure 29
Common symptoms of a heart attack 29
Are NZ foods contaminated by pesticides? 53
Signs of Alzheimer's disease 67
Homemade Yoghurt 87
Chronic fatigue syndrome 113
Immunising vaccines, proven or unproven? ... 148
Should cancer always be treated? 148
Healing advice from Dr Ulric Williams 149
10 minute wholemeal bread 165

"I pass with relief, from the tossing sea of cause and theory, to the firm ground of result and fact."

Winston Churchill

Find me the facts

You may be surprised by the facts contained in this book.

You may also be surprised at the number of embarrassing and miserable health problems we can all avoid and cure by small changes to our diet and a little exercise.

This book has been put together for New Zealanders by a New Zealander. The vital facts you will find in the pages ahead in plain simple language, are the results of the best medical research available up until July 2001. Theories and opinions have been avoided unless supported by years of first hand experience.

You will almost certainly discover from the food tables that your present diet is not giving you the full recommended amounts of some important vitamins and minerals essential for your life-long health. These same food tables show you which foods or supplements you can add to your diet to provide these missing nutrients.

Listed below are some of the questions you will find answered in the book.

Important health questions

- What are the health secrets of extremely long-lived, healthy societies found in some parts of the world?
- Why can some societies (like traditional Eskimos) eat huge amounts of fat but not suffer from heart disease as we do?
- Is dietary cholesterol really harmful? Should I stop eating eggs? Butter? What about Margarine?
- What is the most commonly deficient vitamin in NZ today?
- How can I know if I am deficient?
- Why do some elderly women not suffer from osteoporosis (brittle shrinking bones) almost universal among older women?

- How can I lose weight quickly and keep it off permanently?
- How can I sleep soundly all night without pills?
- What vitamin lack can cause depression?
- How can I avoid the need for tranquillisers?
- What mineral lack can cause prostate swelling, or prostate cancer and impotency in men?
- What mineral lacking in many NZ soils protects against all types of cancer?
- What nutrient can heal and reverse heart disease?
- How can I normalise my high blood pressure?
- How can I prevent brown, old age spots on my skin?
- Can I heal my arthritis in 90 days with good nutrition?
- How can I avoid getting diabetes?
- Does taking antibiotics harm me in any way?
- How can I prevent indoor asthma?
- Is smoking really as harmful as they say?
- What common vitamin deficiency has been strongly linked with miscarriages?
- What vitamin and mineral deficiency has been linked with schizophrenia and other mental disorders?
- Can good nutrition stop my hair turning grey?
- Does taking the pill cause nutrition deficiencies?
- Is fluoride as harmful as they say? What are the true facts?
- What mineral can protect me against deafness in old age?

All these questions and a great many more are answered as this book brings you up-to-date with the latest exciting findings of medical and scientific nutrition research in NZ and around the world.

"Doctor means 'teacher' therefore a doctor's chief duty is to teach people how to be well."

Dr Ulric Williams 1890-1971.
NZ surgeon turned natural healer.

In defence of doctors

Some health books criticise doctors by condemning their lack of training in nutrition. But do the authors really know? Does a doctor stop reading and studying after he graduates?

Most doctors probably know more about nutrition than they are given credit for, but they learn very early in their careers that by far the majority of their patients are not at all interested in changing their eating habits (consider the 98% failure rate of weight loss diets) or in doing much exercise.

Can we blame an overworked doctor if at the end of a consultation he reaches wearily for the prescription pad to prescribe a quick but temporary fix? What would we do in his place?

The average doctor is a highly intelligent and dedicated person, frequently working long hours and receiving little recognition for the stressful nature of his or her work.

However the future is looking brighter. Nowadays more and more doctors, especially in America are treating the majority of their patients with natural nutrients rather than chemical drugs. One successful doctor claims he has reduced his drug prescriptions by 80%.

Nearly all of the respected nutrition works consulted for this book have been authored by former drug treatment doctors, and all of them are enthused by the results they are getting from nutrient therapy and supplements.

For example consider the following disorders:

- Cancers
- Constipation
- Diabetes
- Haemorrhoids
- Heart Disease
- Kidney Stones
- Obesity
- Prostate Troubles
- Varicose Veins
- Gall Bladder problems

All these ailments and many others, all rare in poorer countries, are usually the result of three main causes:
- Wrong diet.
- Mental stress.
- Lack of exercise.

The so called 'western lifestyle.'

For years we have expected our doctors to do the impossible and drug us back to health.

However the information in the pages ahead can help each one of us, with guidance from a sympathetic doctor or natural health practitioner, plus a little self-discipline, to greatly improve and maintain our health.

Choose a health professional who enjoys his or her work, has a warm reassuring smile, a sense of humour and a caring attitude. These qualities are health-promoting in themselves.

Try and choose a doctor who has a warm reassuring smile.

"The first requirement of a good life is to be healthy."
Herbert Spencer 1820-1903. English philosopher.

Layout of the book

We look first at Alcohol, Calories, Carbohydrates, Cholesterol, Fats, Fibre, Protein and Sugar. Then one by one we look at all the important Vitamins and Minerals necessary for our good health, and provide a food table for each one showing their availability in typical servings of NZ foods.

At the top of each food table is the RDI (Recommended Daily Intake) showing how much of that particular vitamin or mineral should be in our diet.

<u>To keep things simple and to avoid clutter, only when a typical serving of food provides 10% or more of our RDI does it appear in the table.</u>

Each of the sections opens with some interesting and useful facts about that particular nutrient.

Then come headings as follows:

Role in our body

Under this heading is listed the role played in our body of that particular nutrient as far as medical science understands it today.

Too little

Here you will find listed the most commonly reported symptoms of a deficiency of that particular vitamin or mineral.

It is extremely unlikely that every symptom would show up in any one person. You will also see that some symptoms are common to deficiencies in other nutrients, low energy for example.

The term 'factor' after a symptom means that a deficiency only contributes to the symptom, but is not believed to be a primary cause.

Too much

Here is listed the known symptoms of an excess dosage. Many nutrients are toxic (poisonous) in high doses. Vitamins A and D for example, and the minerals Copper, Iron and Selenium.

Cooking losses
The effects of cooking if any are listed here. Contrary to popular belief cooking, including microwave cooking, has little effect on most vitamins and virtually none on minerals.

The RDI (Recommended Daily Intake)
The RDI figure you will see at the top of each food table is the recommended daily intake of that particular nutrient to maintain the health of an average adult or child.

The average adult is defined as male weighing 71 kg and female 58 kg, (although NZ male and female average weights have now crept up to 80 kg and 65 kg). However our height rather than our weight is a greater determining factor in our nutrition requirements.

These RDI figures, or RDA as they are termed in America (Recommended Daily Allowance), are set by most major countries and are reviewed every few years as more knowledge becomes available.

As NZ has not set RDI's, the ones listed in this book are Australian and were last revised in 1991. There is not a big difference from country to country but Australian RDI's tend to be on the low side by international standards.

Where no Australian RDI has been set, an American RDA has been used or a typical figure averaged from nutritional sources.

RDI's are intended to suit 97% of the population, and because we all vary in body size and ability to absorb vitamins and minerals they are set on the cautious side. Two thirds of the RDI should be sufficient for 84% of the population.

Some of us may require only as little as half the RDI figure, but many factors can hinder the absorption of nutrients, especially poor digestion due to stress. Also smoking, caffeine, the birth control pill and many pharmaceutical drugs.

Some nutrients have special RDI's for children but most are calculated on a weight pro-rata scale, ie a 29kg girl (half adult weight) would be 50% of the adult RDI.

Pregnant and breast feeding women have higher requirements than normal, around 20%, but this is mostly taken care of by increased appetite. However the intake of Folate (Folic Acid), Vitamin C, Iron and Zinc should be doubled during pregnancy.

Final sections

In the **Conclusions** chapter we attempt to draw some, from the ground covered.

Next is a section on **Becoming Permanently Slim**, the quickest and most effective and even enjoyable way to this elusive state.

Then for those who may be interested, a quick look at some very impressive health statistics from an American study of 6000 Latter Day Saints (Mormons) who along with Seventh Day Adventists live diet laws that promote excellent health.

The next section lists all the mysterious additive Code Numbers that we see on processed food packages, and alongside each one an explanation of what they really are, and whether any adverse reactions have been reported.

Finally there is a comprehensive index to help you quickly locate information in the book.

Higher RDI requirements for pregnant women are mostly taken care of by increased appetite.

"A full stomach does not like to think."

German Proverb

Measures & abbreviations

Listed below are measures and abbreviations you will find in the food tables and elsewhere in the book:

mcg	microgram or µg (one thousandth of a milligram).
mg	milligram (one thousandth of a gram).
gm	gram (one thousandth of a kilogram).
tsp	one level teaspoon (5 millilitres ml).
tbsp	one level tablespoon (15 millilitres ml).
cup	250 mls or 1½ tea cups.
cals	kilocalories (4.2 kilojoules).
med	medium size.
svg	typical serving. *
est	estimated.

* Typical servings are defined as follows:
1 medium size fruit or potato or 2 small fruits.   ½ cup of diced fruit or vegetables.
1 cup of leafy vegetables.   ¾ cup of fluid.

Food tables

The food tables are made up of typical servings of common foods that provide at least 10% of our daily needs of that particular nutrient. Bear in mind the following points when using the tables.

- Meats can vary in nutritional value according to cut so an average figure has been used. (Liver which is rarely consumed but very rich in some nutrients has not been included in that average.)
- Fish can also vary from species to species, but to keep things simple an average figure has been calculated using the most common varieties, but not including shellfish.
- 100 gms is about 3½ ounces, or a quarter of a pound.
- Yoghurt nutrients are virtually identical to milk and have therefore not been listed separately.

- Nutrient figures given for baked beans, usually Haricot, can generally be applied to other beans such as Lentils, Mung beans, Soya beans and Lima beans.
- Fruit juices listed are pure and unconcentrated.
- The figures for canned fruits include the juice but those for canned vegetables do not.
- Trim type, reduced-fat milks and non-fat milks typically contain about 20% more milk solids than regular milks and are therefore correspondingly richer in some nutrients.
- Many breads sold as wholemeal contain approximately 50% wholemeal and 50% white flour. Figures quoted for wholemeal bread are for 100% wholemeal.
- One bun on average is equivalent to two slices of bread.

"It appears to me necessary for every physician to strive to know what man is in relation to his articles of food and drink, and what are the effects of each of them, for by every one a man is affected and changed, and the whole of his life is subjected to them."
Hippocrates 460-377 BC. 'The Father of Medicine.'

A standard measuring cup (250 ml or 8 oz) and contents.

"I have four good reasons for being an abstainer: My head is clearer, my health is better, my heart is lighter, and my purse is heavier."

Thomas Guthrie 1860. (Scottish clergyman).

Alcohol

Pure alcohol is a spirit similar to petrol, they run racing cars on it, yet our body is still able to convert it into energy. Two tablespoons of alcohol contain about 70 calories, so it can be weight forming as many beer-bellied New Zealand males amply demonstrate.

Alcohol is similar to sugar – both are foods high in calories but barren of nutrients, and therefore displace nutritious food that we would otherwise eat.

Too much alcohol hinders the absorption of some important vitamins and minerals, particularly Folate, Pantothenic Acid, Magnesium and Zinc. As a result, serious malnutrition is more common among alcoholics than any other group of people in the western world.

Around 90% of alcoholics are deficient in Folate, a vitamin that contributes to a calm mind and smooth muscle action. The increased tension and anxiety that results can make it all the more difficult to break the addiction.

As most alcoholics drink to allay anxiety, many researchers believe there is a cause and effect link between nutritional deficiency and alcoholism – especially the B group vitamins which play a key role in maintaining our emotional stability.

In one hospital study, 60% of alcoholics benefited from Niacin (Vitamin B3) supplements, and 20% maintained complete sobriety afterwards, compared to 0% with conventional treatment.

The muscle tremors or 'shakes' of the chronic alcoholic are usually caused by lack of Niacin or Magnesium.

Although alcohol can be very toxic to our body there appears to be no adverse health effects from light drinking. In fact some positive health benefits have been reported from a single beer or wine a day, probably due to the anxiety-dulling effects of alcohol, and anti-oxidant activity from the fruits used in making the drink.

"Everything in excess is opposed by nature."
Hippocrates

Calories

Calories are a measure of the amount of energy our bodies can extract from food – one calorie will raise the temperature of one litre of water 1°C.

For purists, the correct term is kilocalories (1000 calories) but common usage has shortened it to calories, or the abbreviation cals. A rough conversion to the metric kilojoules (kJ) is to multiply by four, ie 100 cals equals approx 400 kJ.

Listed on the next page are the typical calorie needs of active New Zealanders according to age, sex and height.

Inactivity or summer warmth decreases our need and winter cold increases our need. Severe cold can treble our requirement – 8000 calories daily is not unusual for polar explorers.

Despite calorie tables, our own appetite and weight level best determine our calorie needs. Provided we are maintaining normal weight, not feeling the cold unduly, and eating <u>just sufficient</u> to satisfy our appetite, our calorie intake should be correct, no matter how low it may be.

Calorie intake and health

Whenever our calorie intake exceeds our needs, we open the door to a host of unpleasant health disorders which can include the following:

- Allergies
- Obesity
- Heart Disease
- Anxiety
- Insomnia
- Diabetes
- Respiratory Infections

A medical phenomenon occurred in Europe during the close of World War 2, following widespread starvation. Medical professionals noted a dramatic fall-off in the incidence of all the above listed disorders, including Cancer.

Also, despite sleeping in cold, wet outdoor conditions for weeks at a time, servicemen reported a complete absence of colds among whole divisions, <u>whilst on restricted rations.</u>

Average daily calorie needs

	Age	Male	Female
Babies up to	1 year	100 cals per kg	
Children	1-2 years	1200	1150
	3-5 years	1700	1500
	6-8 years	1950	1750
	9-11 years	2200	1900
Adolescents	12-14 years	2500	2100
	15-17 years	2900	2200 *

	Height	Male	Female
	160cm / 5ft 3"	2300	2000 *
Adults	170cm / 5ft 7"	2400	2050 *
18-65 years	180cm / 5ft 11"	2500	2150 *
	190cm / 6ft 3"	2700	2250 *
	200cm / 6ft 7"	2900	

* Pregnant women should increase the above +10%, and when breast feeding +30%.

		Male	Female
	160cm / 5ft 3"	2000	1600
Adults	170cm / 5ft 7"	2100	1650
over 65 years	180cm / 5ft 11"	2200	1700
	190cm / 6ft 3"	2400	1800
	200cm / 6ft 7"	2550	

Typical calorie usage Calories per hour

	Men	Women
Lazing / Sleeping	60	45
Light work	100	75
Brisk walking / Cycling	300	225
Running	1200	800

NZ nutrition survey results

The 1997 NZ nutrition survey found the average reported calorie intake for NZ men under 65 was a high 3000 cals (average height 177cm / 5ft 10") and for women 2000 cals (average height 164cm / 5ft 5").

The men's intake was 20% above daily average needs, which would account for their 15% increase in weight since the last survey eight years previously.

However the women, who claimed to be consuming about the right number of calories, had in fact increased in weight even more than the men – by 19%. It is suspected that a few chocolate biscuits may have gone unreported in the survey.

The main source of calories for New Zealanders was reported to be bread.

Depressing statistic

To burn off the calories of a medium size, 120 gm bar of chocolate (630 cals) we would need to:
- Laze for 12 hours.
- Work moderately for 6 hours.
- Walk briskly or cycle for 2½ hours.
- Run or swim for ¾ hour.

HEALTH HINT   The first law of health

Never eat when not hungry

Sit down to at least one meal a day hungry.
Skip a meal if not hungry.
Avoid eating between meals.

"Dine with little, sup with less, do better still, sleep supperless."
Thomas A. Edison

Calories (To convert to kJ's, multiply by 4.2)

BEVERAGES

		Cals
1 can	Beer	110
1 bottle	Beer (small)	105
1 mug	Beer	95
1 jug	Beer	310
1 glass	Wine	70
1 nip	Spirits	40
1 glass	Fruit Juice (pure)	80
1 glass	Fruit Cordial	90
1 can	Cola / Fizzy Drink	160
1 can	Cola / Fizzy Drink (diet)	5
1 cup	Milo (all milk)	215
1 cup	Milo (¼ milk)	100
1 cup	Tea/Coffee (milk & sugar)	40

DAIRY / EGGS / FATS

1 tbsp	Butter/Margarine	110
1 slice	Cheese (4mm)	60
2 tbsp	Cream	110
½ cup	Ice Cream	120
1 lge	Ice Cream cone	120
1 cup	Milk (standard)	160
1 cup	Milk (low-fat)	110
1 cup	Milk (non-fat)	90
1 cup	Custard	280
1 pot	Yoghurt sweetened	150
1 pot	Yoghurt natural low fat	75
1 med	Egg (boiled)	75
1 med	Egg (fried)	120
1 tbsp	Beef Dripping	110
1 tbsp	Vegetable Oil	125
2 tbsp	Salad Dressing	100

FRUITS

1 med	Apple	60
1	Apricot	30
½	Avocado	190
1 med	Banana	100
½ cup	Berry Fruit	40
5	Dates	100
1	Feijoa	15
½ cup	Fruit Salad (canned)	125
10	Grapes	35
1	Grapefruit	70
1	Kiwifruit	50
1	Nectarine	60
1 med	Orange	40
1 cup	Pawpaw	125
1	Peach	50
½ cup	Peaches (canned)	100
1	Pear	70
1	Persimmon	65
½ cup	Pears (canned)	100
½ cup	Pineapple (canned)	100
1 med	Plum	30
5	Prunes	90
1/3 cup	Raisins / Sultanas	130
½ cup	Rhubarb (stewed/sugar)	100
5	Strawberries	35
1	Tamarillo	20
1/8	Watermelon	50

GRAINS

½ cup	Flour	205
1 slice	Bread (thin)	50
1 slice	Bread (med.)	70
1 slice	Bread (med buttered)	170
1 slice	Bread (thick)	90
1	Bread Roll	140
1	Sandwich (avg)	190
1	Bun (cream)	190
1	Bun (current)	240
1	Scone (buttered)	230
1	Doughnut (plain)	170
1	Eclair	160
1	Crispbread (rye)	20
1	Biscuit (small plain)	40
1	Biscuit (choc coated)	60
1	Biscuit (icing filled)	60
1	Shortbread	65
1	Biscuit (home baked)	90
1 svg	Cake (plain)	160
1 svg	Cake (fruit)	165
1 svg	Cake (sponge with jam)	180
1 svg	Cake (iced)	300
1 svg	Cake (rich, iced)	440
1 svg	Steam Pudding	235
1 svg	Christmas Pudding	390
1 svg	Cheese Cake (190 gms)	770
1 svg	Trifle	430
1	Pikelet (med)	60
1	Pancake	110
½ cup	Rice (cooked)	130
½ cup	Rice Pudding	170
¾ cup	Spaghetti (canned)	130

NUTS

1/3 cup	Peanuts (salted oiled)	300
1 tbsp	Peanut Butter	105
1/3 cup	Other nuts (avg)	280
½ cup	Coconut flesh	165
1 tbsp	Coconut shredded	40

	MEATS (cooked)	Cals
100 gms	Beef (grilled or stewed)	200
100 gms	Beef (corned)	220
100 gms	Beef (roasted)	222
100 gms	Chicken (baked)	160
100 gms	Chicken (baked w/skin)	220
100 gms	Lamb (grilled)	220
100 gms	Lamb (roasted)	265
2 slices	Bacon (fried)	140
100 gms	Pork (grilled)	250
100 gms	Pork (roasted))	280
1	Sausage (grilled)	225
3	Sausage Rolls (med)	630
3	Meat Pasties	420
100 gms	Mince	210
100 gms	Lasagne	150
2 slices	Luncheon	115
2 slices	Meatloaf	120
2 tbsp	Gravy	30
	FAST FOODS	
100 gms	Chicken (fried KFC)	310
1	Hamburger	250
1	Hamburger (Big Mac)	550
1	Eggburger	520
1	Cheeseburger	500
1	Hot Dog	430
100 gms	Fish (fried in batter)	300
1 cup	Potato Chips (fried)	115
1 svg	Pizza	225
1 cup	Potato Salad	240
½ cup	Coleslaw	85
1	Meat Pie (small)	350
1	Fruit Pie	240
1	Milkshake	290
1	Ice Cream Sundae	260
1 bag	Potato crisps (med bag)	260
1 bag	Potato crisps (small bag)	85
	FISH	
100 gms	Fish (baked)	80
3	Fish cakes	240
100 gms	Fish (canned)	220
6	Oysters (battered)	310
6	Oysters (raw)	80
1 tbsp	Fish Paste	10
	VEGETABLES (cooked)	
¾ cup	Baked Beans	130
5 slices	Beetroot (canned)	65
1 med	Carrot (raw)	15
5 slices	Cucumber	5
½ cup	Green Beans	10
1 med	Kumara	120
½ cup	Lentils	95
3 leaves	Lettuce (raw)	5
6 med	Mushrooms (microwaved)	10
6 med	Mushrooms (fried)	160
1	Onion	40
½ cup	Peas	45
2 med	Potatoes (boiled)	170
1 cup	Potato (mashed/butter)	200
¾ cup	Pumpkin	35
½ cup	Silverbeet	10
1 cup	Soup (avg)	110
½ cup	Sprouts (raw)	30
¾ cup	Sweet Corn (canned)	105
2 cobs	Sweet Corn	310
1 med	Tomato (raw)	20
¾ cup	Vegetables (mixed)	50
½ cup	Yams	85
	SWEETS	
1 tsp	White Sugar (raised)	30
1 tsp	Raw Sugar (raised)	25
1 tbsp	Golden Syrup	60
1 tbsp	Honey	60
1 tbsp	Jam	50
1	Meringue (with cream)	170
1 tbsp	Milo/Ovaltine dry	30
2	Chocolates	40
1 small	Chocolate bar (50gm)	260
1 med	Chocolate bar (120gm)	630
1	Mars bar	265
1	Muesli bar	130
2	Peppermints	30
2	Toffee Lollies	45
1 cup	Pop Corn (candied)	40
1 cup	Jelly	180
1	Ice Block	60
	HEALTH FOODS	
1 tbsp	Brewers Yeast	10
100 gms	Liver (stewed)	200
1 tsp	Marmite/Vegemite	10
2 full tsp	Malt extract	90
1 tbsp	Molasses	50
½ cup	Tofu	110
2 tbsp	Wheat Germ	60
2 tbsp	Wheat Bran	10
	BREAKFAST CEREALS	
	(With milk and sugar)	
1 plate	Muesli	250
2	Weetbix	200
1 plate	Porridge	200
1 plate	Wheatflakes	200
1 plate	Rice Bubbles	150

Carbohydrates (Complex)

Complex carbohydrates are:
- Grains (flour, bread, rice, cereals, biscuits, etc)
- Nuts • Fruits • Vegetables

Sugars are also carbohydrates but these are called simple carbohydrates and we look at them later on.

Complex carbohydrates need to pass into our intestines and be digested before entering the bloodstream. This takes about two to four hours to complete (unlike refined sugar and alcohol which can enter direct from the stomach) therefore they release energy and nutrients at a steady rate over a long period of time.

Complex carbohydrates especially fruits also contain sugar, but need to be mostly digested before our body can obtain the sugar.

Compared to simple carbohydrates (sugars), complex carbohydrates are a rich source of nutrients. See below how 400 calories of potato, a complex carbohydrate compares with 400 calories of honey, a simple carbohydrate sugar but generally regarded as nutritious.

	400 cals of Potato	400 cals of Honey
Protein	10 gms	1 gm
Vitamin C	45 mgs	0 mgs
Calcium	20 mgs	9 mgs
Potassium	1700 mgs	88 mgs

The pattern is similar throughout the nutrient range.

Another advantage of complex carbohydrates is their generally high fibre content. High fibre foods usually give us a full feeling in our stomachs long before we have eaten excess calories. For example try eating six large carrots in one sitting, which total less calories than just one sausage.

Sports research and field testing of athletes have proven beyond doubt that <u>unrefined</u> complex carbohydrates, (along with plenty of water) is the ideal food for sustained energy. It had long been believed that proteins such as meat and eggs were the best foods for long-lasting endurance.

An interesting list of foods has recently been compiled for athletes ranking the speeds from 0 to 100 at which different carbohydrate foods are converted to glucose for the body to use. It is called the Glycaemic Index. The results are surprising. Grains and vegetables rank high and fruits low.

The RDI of carbohydrates

According to strict nutritionists, the ideal ratio of fats, proteins and carbohydrates in our diet is as follows:

Complex Carbohydrate	75% of daily cals
Simple Carbohydrate (Sugars)	0% of daily cals
Protein	10% of daily cals
Fat	15% of daily cals

However such a diet is unpalatable for the majority of us, so the following compromise guidelines have been set by the NZ Nutrition Taskforce:

Complex Carbohydrate	40% or more of daily cals
Simple Carbohydrate (Sugars)	15% or less of daily cals
Protein	15% of daily cals
Fat	30% or less of daily cals

The 1997 NZ nutrition survey revealed the following actual intakes:

Complex Carbohydrate	34% of daily cals
Simple Carbohydrate (Sugars)	14% of daily cals
Protein	15% of daily cals
Fat	37% of daily cals

So at present we are still eating too much fat and not enough complex carbohydrate. This probably largely accounts for the steady weight increase of New Zealanders, and the widespread incidence of diabetes.

"Man lives on one quarter of what he eats. On the other three quarters lives his doctor."   Old Arab proverb.

"A man is as old as his arteries."   Traditional proverb.

Cholesterol (and artery disease)

Cholesterol has the image of a rampaging killer in the minds of most New Zealanders. This is understandable – it is a mix of cholesterol (a waxy substance) and Calcium that narrows and hardens our arteries, especially the coronary arteries that supply blood to our heart muscle.

Current odds are that a narrowed artery, blocked by a blood clot will be the cause of our own death.

Common belief is that our diet is to blame for this narrowing, especially fatty foods high in cholesterol such as eggs.

The whole truth is much more complex than this however, and from the massive amount of research that has taken place over past years some very interesting facts have emerged.

Fact 1

There are two main types of cholesterol found in our bodies – high density (HDL) which is good and essential for many functions in our body, and the level of which usually remains constant. And low density which is bad, and helps clogs our arteries and gall bladders and also raises the fat levels in our blood. This low-density cholesterol comes in three forms, LDL, LA and VLDL (bad, very bad and even worse) and the levels can fluctuate wildly.

Contrary to general belief, both types of cholesterol are manufactured by our bodies in the liver, not absorbed direct from our food. However although the ratio of good to bad is determined by the types of food we eat, it is also greatly influenced by our stress levels and the amount of exercise we get.

Fact 2

60% of people with clogged arteries <u>do not have</u> high levels of harmful low density cholesterol in their blood.

Fact 3

A recent 10 year study of 374 young men and women involving regular heart scans, found that the 17% who scored above average on a psychological test for hostility, accumulated over twice as much Calcium deposit in their arteries (an indicator of heart disease) than those who scored below average in the test.

Fact 4
Animals fed a diet of high cholesterol foods sometimes develop clogged arteries, but not humans – provided adequate Vitamin C, Chromium, Selenium and Zinc levels are maintained.

Fact 5
Traditional meat and fish eating Eskimos who are high dietary cholesterol consumers, seldom develop hardened and clogged arteries. Also a 2001 study found that women who consume at least 225 gms of fish a week cut their risk of stroke 47%.

Fact 6
Prolonged high blood pressure can erode the smooth inner linings of our arteries, especially at the forks, providing a key for cholesterol and Calcium deposits to attach and build up. These deposits eventually narrow and harden our arteries, restricting the flow of blood and further increasing our blood pressure.

When an artery blocks in a crucial place, the result is a heart attack or stroke. This blockage is usually caused by a blood clot.

Fact 7
High blood pressure can be caused by fat-thickened blood, just as the oil pressure of a car is higher when the engine is cold and the oil thick. It can also be caused by emotions such as fear, anger and anxiety that temporarily narrow our arteries.

Fact 8
Animal fats such as those found in meat, butter and eggs (saturated fats) can increase the amount of harmful blood cholesterol manufactured by our body.

Fish oils, vegetable oils and plant-sourced cooking oils (unsaturated fats) can lower the amount of harmful blood cholesterol manufactured. However if any of these oils have been artificially hardened such as margarine, their effect is similar to animal fats. Both types of fat temporarily thicken our blood during digestion.

Fact 9
Stress, smoking, excess sugar, alcohol and caffeine, overeating

or lack of exercise will generally raise cholesterol levels in males over 14 years of age and females over 45 years.

Fact 10
In a USA study, men were fed three eggs daily. Cholesterol levels increased 27% in the smokers, but only 2% in the non-smokers. Smoking hinders the absorption of Zinc and destroys Vitamin C, both of which are crucial to maintaining low cholesterol levels.

Fact 11
Lack of the following nutrients can raise cholesterol levels:
- Chromium
- Manganese
- CoQ10
- Niacin
- Fibre
- Selenium
- Inositol
- Vitamin C
- Iodine
- Zinc
- Magnesium

Fact 12
Too much of the following can also raise cholesterol levels:
- Alcohol
- Cobalt
- Caffeine
- Copper
- Cadmium
- Sugar
- Calories

Fact 13
Being male, obese <u>and unfit</u> increases the risk of death through artery disease by 60%. However being obese <u>and fit</u> reduces the risk to normal.

Fact 14
There is a worldwide incidence of higher cholesterol levels in areas where the drinking water is soft. Soft water is mildly acid and can absorb excessive amounts of the potentially harmful minerals Copper and Cadmium from water pipes which eventually enter the body.

Also soft water also does not normally contain Magnesium which assists in maintaining safe levels of cholesterol.

Fact 15
Human vegetarians and apes, whose food is very low in cholesterol, often develop clogged and hardened arteries in <u>prolonged, stressful situations.</u>

A very significant American study found substantially higher blood levels of cholesterol in shift workers who encountered different co-workers during each shift, as compared to those who worked with familiar faces each shift – an obvious stress factor.

A further English study of 1600 workers whose jobs were at risk over four years, found a 50% increase in heart disease along with a similar increase in other health and social problems.

Fact 16
Slow paced, family orientated, vegetarian societies have the lowest cholesterol rates in the world, and also the longest life spans. Average total blood cholesterol levels of some of these societies are below 3.0 mmol/L compared with the New Zealand average of over 5.7. A healthy level is regarded as 4.5 and dangerous over 6.5.

Conclusions
Taking into account all of the above facts and others reported later in this book leads us to several logical conclusions.

1. The main cause of excess cholesterol levels, which can lead to clogged arteries, high blood pressure, heart attacks and strokes, and also according to research during 1999, Alzheimer's disease, is <u>prolonged negative stress</u>, aggravated by smoking, over-eating, lack of exercise and excess caffeine.

2. Contributing dietary factors which can increase cholesterol levels by aggravating emotional stress are:

- Excess Caffeine.
- Excess Copper.
- Excess Cadmium.
- Lack of Niacin.
- Lack of Inositol.
- Lack of Folate.

3. Dietary factors that can hinder our body in the manufacture and maintenance of safe levels of cholesterol are:

- Lack of Chromium.
- Lack of Fibre.
- Lack of Iodine.
- Lack of Magnesium.
- Lack of Manganese.
- Lack of Selenium.
- Lack of Unsaturated Fat.
- Lack of Vitamin C.
- Lack of Zinc.
- Lack of fish oil.
- Excess Cobalt.
- Excess Saturated Fat.

Avoiding heart disease

Eating less high cholesterol foods can lower our cholesterol levels to a small degree, but the best place to tackle the heart disease problem is at a more effective level such as:

- A contented lifestyle (a peaceful marriage helps).
- Adequate exercise (see section on becoming permanently slim).
- Not smoking.
- A diet of whole grains, fruits and vegetables, limited in sugar.
- Just sufficient calorie intake.
- Minimal use of caffeine and alcohol.

We should especially ensure that our diet contains adequate CoQ10, Vitamin C, Selenium and Zinc, four very important anti-oxidant nutrients that protect our cells and assist our body in expelling harmful toxins. Also Pectin, a substance found in many fruits especially apples also helps limit the amount of harmful cholesterol our body manufactures.

Also sufficient Inositol and Niacin to help us cope with the stresses of life.

Finally a word on dietary fat, which many believe is the main cause of heart disease. Fat consumption in the western world has not increased significantly during the past 90 years, but during that time the incidence of heart disease has increased enormously.

What <u>has increased significantly</u> during that same period is, the consumption of sugar and refined flour, the use of motor vehicles (resulting in much less walking), and stress, due largely to the breakdown of the traditional family and uncertainty of employment.

"Eating foods containing cholesterol have never been proved implicit in the cause of heart disease."   Dr G. Meinig. DAS. FACD.

"Saturated fats are not the cause of coronary heart disease. That myth is the greatest scientific deception of this century."
      Dr George G. Mann. MD Author *"The Cholesterol Conspiracy"*

"33 clinical trials have failed to show that cholesterol and saturated fat intake, or cholesterol lowering medication protect against coronary heart disease."   Dr Robert Crayhan.

Blood pressure guidelines

Men over 35 and women over 45 should have their blood pressure checked every few years.

Systolic or pumping pressure, the higher of the two is normally around 120 in a healthy adult and the lower diastolic pressure about 70, normally expressed as 120 / 70.

130 / 80 is average for a 40 year old NZ male, and 117 / 75 for a 40 year old female. Ideally our blood pressure should not increase with age. In long-lived, low cholesterol societies it remains stable, but in Western societies both cholesterol and blood pressure steadily increase with age.

A systolic pressure of 160 or above is regarded as dangerously high, as also is a diastolic over 90. However abnormally high blood pressure should be rechecked a week later as results can be influenced up to 50% by temporary factors. 20% of New Zealanders suffer from high blood pressure.

If the problem persists, an electrocardiogram incorporating a treadmill test can provide a reasonable picture of heart and artery condition, but is not 100% reliable.

For purposes of measuring longevity, the lower diastolic or resting pressure is the important figure. Listed below are the average number of years lost from a man's life by disorders brought on by high diastolic blood pressure at age 45. Average is about 80.

Diastolic Pressure	Average years lost at age 45
90	3 years
95	6 years
100	12 years

Cholesterol guidelines

Cholesterol levels can be checked at the same time as blood pressure. Ideally they should be around 4.0 mmol/L or lower, but this is rare in western societies.

The NZ average is 4.5 for teenagers steadily increasing to 6.5 for 70 year olds. Average is 5.7. Over 5.0 is generally regarded as unsafe and over 7.0 dangerous.

Anxiety induced high blood pressure

In handling the stresses of life, people can usually be divided into two basic types – Confronters and Avoiders. Confronters face up to life's problems while Avoiders ignore them hoping they will go away.

In an experiment on anxiety induced high blood pressure, a large group of volunteers were given periodic sharp electric shocks at timed intervals. The Confronters in the group watched the clock, braced for the shock, then relaxed again. The Avoiders did not watch the clock, not wanting to know when the shock was coming.

As a result, due to their constant tension, the Avoider's blood pressure was continually an average of 30% higher than the Confronters, all of whose blood pressure remained normal.

Common symptoms of a heart attack

1. Breathlessness.
2. Nausea.
3. Sweating.
4. Crushing sensation in the chest.
5. Pain in chest, arms, jaw, or neck.

In one third of heart attacks, symptom number five is absent, ie, there is no pain. If any of the other symptoms appear with no obvious cause, seek medical help quickly.

HEALTH HINT   Smoking and cholesterol

Smokers over the age of 40 are twelve times more likely to have high levels of harmful blood cholesterol than non-smokers the same age.

However after quitting for 10 years the risk of heart disease and stroke drops to that of a non-smoker.

"For a modern disease to be related to an old fashioned food is ludicrous."   Dr T. L. Cleave.

Fats (also known as Oils or Lipids)

The weight-watcher's arch enemy, gram for gram fats contain over twice as many calories as carbohydrates.

For example one medium thick slice of bread contains about 70 calories, but by spreading on a level tablespoon of butter or margarine we add a further 110 calories.

More examples are listed below, all based on an identical 100 gram weight of food.

100 gms Carbohydrate Food	Calories	100 gms High Fat Food	Calories
Cabbage	25	Meat (red)	200
Banana	80	Chocolate	530
Potato	85	Margarine	740
Bread	220	Cooking Fat	890

However, as fat is highly concentrated it is slow to digest and therefore prevents us from feeling hungry again for a longer time than any other type of food.

Fats and disease

Many serious diseases are blamed on high fat consumption, yet traditional Eskimos consume huge amounts of fat with no sign of the high blood pressure and heart disease that trouble the western world.

Part of the explanation is that the Eskimos burn up fat keeping warm in their cold harsh climate, whereas in mild climates and sedentary lifestyles such as ours, the calories are surplus.

Supporting this view is the fact that our fat consumption has not increased in recent years, but fat-linked diseases have increased greatly. It is becoming apparent that it is not so much the fat that is harmful, but the excess calories due to lack of exercise, especially walking, plus stress, smoking, and lack of certain key vitamins and minerals due to over processed and refined foods.

Another factor in Eskimo health is that much of the fat they consume comes from fish. Fish oil tends to be richer in the

Omega 3 type of unsaturated fat, which has natural blood thinning qualities and therefore protects against blood clots. Blood clots are a major factor in heart disease.

Cod Liver Oil, Sardines and Flax Oil are also high in Omega 3.

If our arteries have narrowed, the risk of a heart attack or stroke is highest during the hours following a rich fatty meal. Blood thickened by fat particles during digestion is more subject to clotting.

Excess coffee can increase the danger. Dutch researchers gave volunteers six cups of strong coffee a day for two weeks. At the end of that time they found that their blood fat levels had increased 36%, also their Vitamin B6 levels had dropped 21%.

Excess fat, or rather the excess calories from fat (1 kg of fat contains 9000 calories), is believed to be the main cause of the ever increasing incidence of late-onset diabetes, a very serious health disorder that is doubling every 15 years.

The Pritikin Program

Impressive results in the treatment of advanced heart disease have been claimed by the Pritikin Research Institute in America.

Patients in treatment are given regular small meals of complex carbohydrate food, very low in fats and sugars (and consequently calories). Combined with this diet is a rigorous daily exercise program, all in a low stress supportive environment.

Elderly heart attack victims are reported to recover within a few months from being able to only shuffle a few feet, to walking ten miles a day, and then jogging – all without any drugs whatsoever.

Food is meant to be enjoyed

A certain amount of fat is desirable to enhance the taste and texture of our food. However in New Zealand this has been for years at the excessive level of 40% or more of total calories. This high level is slowly beginning to fall as more people become aware of the health hazards of excessive fat consumption, but was still at 37% in 1997.

Our safest course appears to be to keep our total fat calories below the 30% guideline and closer to 15%, and to try and achieve a two to one balance of unsaturated to saturated fat. (See next section.)

The table below shows the percent of fat in various NZ foods and can give you some idea of your present consumption of fat.

High fat foods

RDI (for fat content of daily calories.) Not more than 30%.

		Percent fat	Total Calories
1 tbsp	Cooking Fat or Oil	99%	125
3 tbsp	Butter/Margarine	82%	330
3 tbsp	Salad Dressing	79%	150
1/3 cup	Peanuts	48%	300
3 tbsp	Cream	40%	165
25 gms	Cheese	35%	105
50 gms	Pastry (flaky)	34%	400
1 med bag	Potato Crisps	33%	260
1 svg	Cheesecake	32%	780
100 gms	Chocolate	30%	530
2 med	Biscuits (chocolate)	28%	120
½	Avocado	26%	190
2	Sausages/Saveloys	25%	450
2 med	Eggs (fried)	23%	240
100 gms	Chicken (fried)	22%	310
100 gms	Fish (fried in batter)	20%	300
2 med	Biscuits (plain)	20%	80
1 svg	Fruit Cake (rich)	16%	440
1 cup	Potato Chips (fried)	14%	115
100 gms	Pork (grilled)	14%	250
100 gms	Lamb (grilled)	12%	220
100 gms	Chicken (baked)	11%	160
2 med	Eggs (boiled)	11%	150
½ cup	Ice Cream	11%	120
100 gms	Beef (grilled)	9%	200
1	Doughnut	9%	170

Typical fat percentages of low-fat foods

Milk (whole)	4%	Fruits	.5%
Bread - cereals	2%	Vegetables	.5%
Fish (baked)	2%	Beverages	0%
Milk (low-fat)	1.5%	Milk (non-fat)	0%
Beans	1%	Sugars	0%

Saturated and unsaturated fats (Fatty acids)

Saturated fat is generally firm, like butter and meat fat and is mostly derived from animals, whereas unsaturated fat is usually liquid at room temperature, ie an oil, and is mostly derived from plants such as Soya beans.

When an unsaturated oil is chemically hardened to a firm fat such as margarine, it takes on the same characteristics as saturated fat for diet purposes.

There are many chemical differences between the two types of fat, yet all natural fats combine both types of fat to a degree, ie, most saturated fats contain about 45% unsaturated fat and most unsaturated fats contain about 15% saturated fat.

The main diet benefits of unsaturated fat are:
- It is easier to digest.
- It tends to reduce harmful blood cholesterol levels.

Unsaturated fats become rancid quickly when exposed to air, but anti-oxidants such as Vitamin A, Vitamin E and the minerals Zinc and Selenium slow this down. Also when this type of fat is heated during cooking or processing, particles called free radicals can form which are harmful to our arteries if they are not destroyed by our immune system. Anti-oxidants in our diet play an important role in this protection.

About a third of most unsaturated fat is also polyunsaturated, but for dietary purposes this is not greatly significant. More significant is the proportion of Omega 3 in the fat, which is one of the three main fatty acids – Omega 3, Omega 6 and Omega 9 that make up all unsaturated fat. Omega 3 has blood thinning qualities and can protect against dangerous blood clotting the main cause of heart attacks. Flax oil, natural fish oils, and Cod Liver oil are high in Omega 3. Soya Bean and Canola oils also contain useful amounts.

Until recent years those at risk from clogged arteries or having high blood cholesterol were advised to consume only unsaturated fat and steer clear of saturated fat which tends to be high in cholesterol, ie butter. However with the discovery of oxidising free radicals which have been linked with cell damage and the calcification (hardening) of arteries, many health professionals about-faced and labeled unsaturated fat as the greater villain.

As is often the case with conflicting claims, the truth justifies both positions but removes the problem. Both types of fat can be harmful in an unbalanced diet, however provided our diet contains sufficient Fibre, along with the anti-oxidant Vitamins C and E, and the minerals Zinc and Selenium, we can handle both types of fat with no apparent harm. Of course our consumption of fat should still not exceed the recommended maximum of 30% of our total calories.

Both fats have roles to play in our body. Current research recommends a 2 to 1 ratio in favour of unsaturated fat, ie twice as many calories of unsaturated fat as saturated fat. These values are listed in the food chart on page 36. The typical NZ diet currently has a 1 to 1 ratio of unsaturated fat to saturated fat.

Role in our body
- Essential to maintaining health of our body cells and blood.
- Helps reduce excess cholesterol (unsaturated fat only).
- Helps our body absorb Vitamins A and D and the mineral Phosphorus.
- Helps regulate our thyroid and adrenal glands.

Too little (Unsaturated fat only)
- High cholesterol levels. Heart attack. Stroke.
- Internal blood clots.
- Gallstones.
- Enlarged prostate gland.
- Arthritis.
- Hyperactivity in children (ADD).
- Hay fever and asthma (factor).

Too little (Both fats)
- Slow growth in children.
- Dull or dry hair and skin. Hair pulls out easily.

Too much
- Slow blood clotting of wounds (unsaturated only).
- Obesity. Late-onset diabetes. Blindness. Gangrene.
- Clogged arteries (saturated only).
- General ill health due to excess calories.

Cooking losses
No significant losses, but can spoil if left exposed to air and warmth.

HEALTH HINT — **Garlic a natural blood thinner**

Garlic following, or during a fatty meal can minimise any harmful effects and speed our blood's return to normal levels of viscosity. (Odourless tablets are available for those who wish to retain current friendships.)

HEALTH HINT — **Avoiding high blood pressure**

To maintain normal blood pressure we must ensure a balanced intake of approx 50/50 Potassium and Salt.

We must also keep sugar intake below 15% of our calories, ensure a 2 to 1 ratio of unsaturated to saturated fat, maintain the correct RDI of Magnesium, CoQ10 and Calcium, and not exceeding the RDI of Protein which can hinder the absorption of Calcium.

Finally we must be ruthless in reducing excess weight, ideally to the point of eliminating all sign of a spare tyre around our middle.

Emotional factors can also cause high blood pressure, see Health Hint below and box on page 29.

HEALTH HINT — **Lower your stress**

Long term anger and resentment (resentment is just repressed anger) are major causes of ill health, especially heart disease.

Anger and resentment however can only exist with blame. If we refuse to fasten blame on others for our problems, our anger will evaporate as quickly as it arises.

So to enjoy continual health and avoid long term anger we should accept <u>full responsibility</u> for all that happens to us, and not blame others for our problems. Life is after all a learning experience, and we have the freedom to choose for ourselves.

Saturated and unsaturated fat

(Only calories of the fat content of the foods are listed)

> RDI (as percent of daily calories)
> Saturated: no more than 10%
> Unsaturated: no more than 20%

		Saturated Calories	Unsaturated Calories
	DAIRY / EGGS		
3 tbsp	Margarine	60	250 *
3 tbsp	Butter	190	120
2 med	Eggs (fried)	50	120
100 gms	Chocolate	175	105
2 med	Eggs (boiled)	45	80
2 cups	Milk (std)	120	65
3 tbsp	Cream	100	55
25 gms	Cheese	50	25
½ cup	Ice Cream	35	20
	FRUITS		
1/2	Avocado	35	110
	GRAINS / NUTS		
1/3 cup	Peanuts (roasted)	50	170
1 tbsp	Peanut Butter	15	60
	MEATS / FISH (cooked)		
100 gms	Fish (fried in batter)	60	115
100 gms	Pork (grilled)	50	75
100 gms	Chicken (baked)	35	65
100 gms	Lamb (grilled)	85	45
100 gms	Beef (grilled)	40	40
	VEGETABLES		
1 cup	Chips (fried in oil)	40	85
	FATS / OILS		
1 tbsp	Olive Oil	20	110
1 tbsp	Cooking Oil (Soya)	20	100
1 tbsp	Cooking Fat (Beef)	45	55

* Although technically mostly unsaturated, margarine has a similar effect in our body to a predominantly saturated fat such as butter.

"Leave gormandising. Know the grave doth gape for thee thrice wider than for other men." — Shakespeare.

Fibre

Fibre is the part of food that our body cannot absorb. It passes through our digestive system largely unchanged.

All food fibre is not dry and rough in texture like bran which most of us bring to mind when we think of food fibre. Foods such as carrots, oranges and pears are also high in fibre.

Why fibre is important

As fibre passes through our intestines it retains water. This water helps keeps faecal matter soft and adds bulk, which allows for easier bowel action – sometimes too easy as many overseas travellers have found after eating several vegetarian meals in a row.

Adequate fibre was long believed to protect against the highly unpleasant disease of colon cancer, but two large recent studies reported in 1997 and 1999 of over 100,000 men and women found that it makes no difference at all. But they did find that adequate Folate decreases the risk of colon cancer 75%.

Too little fibre is however a major factor in constipation, which affects a majority of New Zealanders over 60 years of age. Constipation often leads to haemorrhoids and sometimes the complication of diverticulitis – a bulging of the colon walls.

Study after study also show that fibre plays a key role in maintaining low cholesterol levels by binding excess cholesterol in the blood, which in turn can lower high blood pressure because the blood is thinner.

Cereals are a good source of fibre, however recent research has found that when cereal bran is eaten alone, it can bind useful minerals such as Calcium, Zinc and Magnesium and hinder them from being absorbed by the body. 50% less absorption was reported in one study, when a heaped tablespoon of bran was taken with every meal by elderly people, further testimony to the truth that the best foods are those in their natural state. Good sources of fibre are fruit, vegetables and wholemeal bread.

Excess natural fibre in our diet does not appear to be harmful once our body adapts. However temporary diarrhoea can occur, or its counterpart constipation, when reverting from a high fibre diet back to a low fibre diet, or vice-versa.

Fibre and refined foods

White flour has on average 75% less fibre than wholemeal flour (also called wholegrain flour) and also far fewer nutrients. Greater awareness of this fact has led to a return to wholemeal breads in recent years.

The main purpose of refining grains is to increase their shelf life. It appears to be an unfortunate fact of life that the more nutritious a food, the faster it spoils.

Role in our body
- Necessary for proper functioning of our digestive system.
- Binder to remove excess fats in our diet.

Too little
- High blood cholesterol. High blood pressure.
- Constipation. Haemorrhoids. Diverticulitis.
- Gallstones.

Too much
No long term harmful effects if consumed as found naturally in food.

Cooking losses
Cooking does not significantly affect fibre.

HEALTH HINT — **Avoiding constipation and piles**

Constipation and its two cousins, haemorrhoids (piles) and varicose veins are often caused by too much sitting, even if we are getting sufficient fibre in our diet.

As a general guide we should spend about a third of our waking hours on our feet <u>and moving.</u> And we should not sit longer than three hours at a time without compensating by exercising for about half an hour, especially our legs and abdomen muscles. Brisk walking is excellent.

Correct breathing is also important. We should breathe use our lower abdomen rather than our upper chest (watch a sleeping child). This produces a continual massaging or palpating effect on our intestines similar to walking.

High fibre foods

RDI 28 gms

	FRUITS	**gms**
1	Orange	7.5
½ cup	Fruit Salad (avg)	5.0
½ cup	Raspberries	5.0
1	Pear	4.8
½ cup	Blackberries	4.5
5	Dates	3.6
1 cup	Pawpaw	3.6
1/3 cup	Sultanas/Raisins	3.5
½ cup	Apricots (dried halves)	3.0
½ cup	Rhubarb (stewed)	3.0
4 slices	Beetroot (canned)	3.0
1	Passionfruit	2.9
1	Persimmon	2.9
1	Peach	2.6
5	Prunes	2.5
1	Nectarine	2.3
1	Guava	2.1
1	Tamarillo	2.0
1 med	Banana	1.9
2	Plums	1.6
1	Kiwifruit	1.6
1 med	Apple	1.5
½	Avocado	1.5
½ cup	Strawberries	1.5
1	Apricot	1.1
1	Feijoa	1.0
	MEATS	
2	Sausages	3.4
	GRAINS / NUTS	
½ cup	Flour- Wholemeal	8.0
4 slices	Bread - Wholemeal	6.4
1/3 cup	Peanuts	4.0
2	Weetbix	3.2
4 slices	Bread - White	2.8
1/3 cup	Other nuts (avg)	2.5
½ cup	Flour- White	2.1
1 plate	Porridge	2.0
	VEGETABLES (cooked)	
2 cobs	Sweet Corn	12.0
¾ cup	Baked Beans	7.0
½ cup	Peas	6.5
¾ cup	Broccoli	5.0
6 med	Mushrooms	4.2
¾ cup	Carrots	3.7
¾ cup	Pumpkin	3.6
2 med	Potatoes	3.6
¾ cup	Cabbage	3.0
1 med	Kumara	2.9
1 cup	Cauliflower	2.6
½ cup	Coleslaw	2.2
3 slices	Beetroot	2.2
¾ cup	Parsnips	1.9
1 med	Tomato	1.6
1	Onion	1.6
	HEALTH FOODS	
2	Bran Bix	7.9
1/3 cup	Wheat Bran	7.5
1 plate	Bran Cereal	6.3
1/3 cup	Oat Bran	6.2
1/3 cup	Wheat Germ	5.2
1 plate	Muesli	4.8
1	Muesli Bar	1.2

"You are what you eat."   Old Proverb

Protein

Protein, a relatively expensive food, is essential for health but tends to be over-indulged in by the wealthy western world, mainly through excessive meat consumption. The 1997 nutrition survey found that NZ men consume 105 gms daily, almost twice the RDI.

Excessive protein has been blamed for almost every disease known to man, and the success rate of cure by long term fasting would tend to support many claims, but there is little hard evidence that excess protein is harmful to our body, apart from its tendency to hinder the absorption of Calcium (see page 92).

<u>Eating more calories than we need, of any mix of food, is beyond argument a major cause of ill health.</u>

Lack of protein is still a common nutritional problem in the undeveloped countries of the world. Rice is a poor provider. One cup of cooked rice yields only 5 gms of protein, whereas the RDI for an adult male is 55 gms.

Protein is made up of 22 different amino acids, but not all are found in the right proportions for the human body in any one food, (an egg comes close). Therefore our liver acts as a store house to recombine the various amino acids in the right proportions as required. Over half of these amino acids can be manufactured by our body if necessary.

Our muscles are largely protein, but increasing the amount of protein in our diet won't build bigger muscles, only exercise can do that. Also Protein is not an efficient energy food – athletes have been proven to perform better on high carbohydrate foods.

Meat, fish and dairy products are the most concentrated sources of protein. For those who prefer not to eat meat or are unable to tolerate dairy products, protein is readily available from plant sources, especially beans, nuts and grains.

Role in our body

- Provides the building material for our body cells.
- Required for muscle building and wound healing.
- Helps to regulate our brain functions.

- Helps to regulate our reproductive system.
- Helps to regulate our immune system.

Too little
- Lack of growth and development in children.
- Muscle wasting in adults. Swelling of tissue (edema).
- Dull hair which can pull out easily.

Too much
- Can restrict our ability to absorb Calcium (over 120 gms daily can cause Calcium loss from our bones).
- Osteoporosis (shrinking, porous and brittle bones).
- Kidney failure.

Cooking losses
No significant losses.

HEALTH HINT   Muscle power from water

As the muscle protein in our bodies requires a lot of water to work efficiently, we should drink plenty of plain water before any strenuous work or athletic performance. Lack of water can reduce muscle efficiency by 25%.

DANIEL IN BABYLON

The king assigned them a daily portion of the rich food which he ate and of the wine which he drank.

But Daniel resolved not to defile himself with the king's rich food or the wine which he drank.

So Daniel said to the king's steward, "Test us for ten days — let us be given pulse (peas, beans, lentils, etc) to eat, and water to drink, then compare our appearance with that of the young men who eat the king's food."

At the end of ten days it was seen that Daniel and his three companions looked healthier and better nourished than any of the young men who ate the king's rich food.

Bible Book of Daniel. 1:4-16.

Protein

RDI

			gms	
Children	1-3 years	16		
	4-7 years	21		
	8-11 years	33		
	12-15 years	47		
	16-18 years	Male 67	Female 57	
Adults	Men	55		
	Women	45	Breast feeding 61	

		gms
	DAIRY / EGGS	
2 cups	Milk (std or low-fat)	17
2 med	Eggs	13
25 gms	Cheese	7
	GRAINS / NUTS	
1/3 cup	Nuts	10
4 slices	Bread - Wholemeal	10
4 slices	Bread - White	8
½ cup	Flour- Wholemeal	8

		gms
1 svg	Pizza	8
½ cup	Flour - White	6
	VEGETABLES (cooked)	
2 cobs	Sweet Corn	10
	MEATS / FISH	
100 gms	Meat - White	28
100 gms	Meat - Red	24
100 gms	Fish	16
100 gms	Meat - Processed	14

HEALTH HINT — Minimise colds and flu

Our most important nutrient is oxygen. Without a constant recharging of the oxygen in our blood by breathing, we would die in four minutes. Our daily requirement of oxygen is measured in kilograms rather than grams or milligrams.

One major reason why colds and flu are more prevalent in winter is because we seal our houses against cold draughts (fresh air) which reduces the level of oxygen in the air. Oxygen is essential for our immune systems.

American Plains Indians claim they never caught colds when they lived in well ventilated teepees (which are open to the air at the top) only when they began living in sealed houses, as did the white man.

If we wish to enjoy good health and minimise colds and throat infections we should ensure that our houses are well ventilated all year round, including bedrooms.

"Nectar is poison if taken to excess."   Hindu proverb

Sugars (Simple carbohydrates)

Sugars contain few if any nutrients, only calories. Even honey, widely regarded as a health food contains only minuscule amounts of vitamins and minerals. Calcium is one of the more predominant of these, but to obtain our RDI of Calcium from honey alone we would need to eat about fifty pots a day.

Sugars such as syrups, honey, refined sugar and high sugar foods such as jams, chocolate and sweet drinks provide us with short-term energy more rapidly than complex carbohydrates such as grains. This is because sugars do not need to be slowly digested like other foods and can pass more quickly into our bloodstream.

All sugars, whether they be raw, refined, or from fruit or honey are converted to glucose by our body. Sugar is the fuel our muscles use, and also our brain, which is why a diabetic can go into a coma quickly when glucose levels become erratic.

Once sugars are converted to glucose our body does not recognise any difference between them. However lack of Chromium in refined white sugar can cause problems (see page 97).

Sugar and disease

Excessive sugar consumption is condemned as harmful by nutritional authorities around the world, without exception.

NZ sugar consumption increased from less than one level teaspoon a day per person during the nineteenth century, to a peak of over half a cup per person a day during the 1970's.

NZ males in their early 20's still consume over half a cup of sugar a day, but over the past 30 years overall NZ consumption has fallen 20%, to just under half a cup.

But this level is still excessive and can play havoc with the blood sugar regulators of our body, laying the foundation for serious health disorders.

Fats are also converted to sugar (glucose) by our bodies and are therefore a concentrated source of sugar. However because fats are slowly digested, they do not stress our sugar regulation systems violently, as do high sugar foods. Eating a high sugar food is a little like pouring petrol directly into an engine instead of allowing the carburettor to meter it precisely, according to demand.

A common disorder resulting from high sugar consumption and one that is often unrecognised is hypoglycaemia – wildly fluctuating blood sugar levels, which tend to cause similar fluctuations in our brain, affecting our moods and energy levels. Hypoglycaemia is often a forerunner of sugar diabetes, a potentially fatal disorder in which our body loses control of blood sugar levels.

Both hypoglycaemia and diabetes are on the increase.

One American survey found a 95% incidence of hypoglycaemia among alcoholics and 65% among schizophrenics. Occurrence was also high among hyperactive (ADD) children, drug addicts, violent criminals, and fully 50% of all American mental patients.

Many researchers consider hypoglycaemia to be the root cause of most anti-social behaviour in the world today. This theory has been tested in prisons by restricting sugar consumption. Results have been indifferent. However when combined with soundly nutritious meals, significantly better behaviour is noticed after three months. One institution reported a 40% drop in breaches of discipline.

It appears that the sugar regulation organ of our body, the pancreas can only take so much abuse before it starts to malfunction. This is usually when the diet calories from sugar reach 25%. Stress and lack of exercise also contribute to hypoglycaemia. Symptoms of hypoglycaemia include the following:

- Craving for sweet foods and drinks.
- Sudden fatigue about two hours after a meal.
- Headaches and blurred vision.
- Fainting. Trembling. Cold hands and feet.
- Panic attacks. Depression. Paranoia.
- Aggressiveness. Anxiety. Irritability.

Tooth decay is another major and very expensive problem caused through excess sugary foods, especially sticky sweets that leave a residue in the hollows of our back teeth. Chocolate is a major offender despite some recent claims otherwise.

Heart disease, including high blood pressure can also be caused by excess sugar especially if it is lacking in Chromium (some researchers say it is the main cause). Diabetes and heart disease are often found together. Our blood tends to be sticky and clot more easily when blood sugar levels are high and our immune system also becomes temporarily depressed.

Kidney stones and gall stones are largely attributed to excess sugar (25% or more of daily calories). High sugar intakes also provide a favourable environment for tongue and mouth ulcers, candida, thrush, yeast infections and increased sensitivity to asthma, hay fever and allergies of all kinds.

Hidden sugar

The high sugar levels in modern processed foods are often underestimated. A few examples:

1 bar	Chocolate (med)	17 teaspoons of sugar.
1 slice	Rich cake	14 teaspoons of sugar.
1 can	Fizzy drink	11 teaspoons of sugar.
1 glass	Fruit cordial	8 teaspoons of sugar.
1 svg	Apple pie	4 teaspoons of sugar.
1 svg	Peaches (canned)	3 teaspoons of sugar.

RDI
No more than 15% of calories

NO FOOD TABLE

HEALTH HINT   Tame that sugar craving

A sweet tooth or sugar craving is mostly an acquired taste. We can usually restore our taste buds to normal by severely restricting our sugar intake for one month. This also enhances our enjoyment of the more subtle flavours of vegetables.

We can generally halve the sugar content of most traditional recipes for cakes, biscuits and desserts without significantly affecting the taste. In fact the taste is often heightened as the more delicate flavours are accentuated.

"When 25% of the diet calories are sugar, one is at risk of diabetes, glucose intolerance, heart attack, social problems, excess Calcium in the urine, gallstones and mineral deficiencies."
American Food and Drug Administration. 1986.

"An ounce of prevention is worth a pound of cure."
Old Proverb

Vitamin A – Retinol & Carotene

Some children went blind in Denmark during the last world war when all butter and margarine was exported. This had been their main source of Vitamin A, an important anti-oxidant that plays a major role in the health of our eyes.

Even today in poorer countries of the world, thousands of people lose their sight each year due to Vitamin A deficiency. Our eyes, skin, bones, teeth, hair and nails can all be noticeably affected by lack of this vitamin.

An American survey found 33% of all children and 26% of the general population deficient. Another survey found Vitamin A levels 85% below normal among a group of cancer patients.

In NZ only 7% of under 25 year olds in poorer communities were found to be potentially deficient in the 1997 nutrition survey, but actual blood levels were not tested.

Prolonged negative stress, especially as the result of injury has been found to drop Vitamin A levels dramatically.

Too much Vitamin A can also cause problems. Back in the 16th century a group of Dutch seamen ate polar bear liver which is extremely rich in this vitamin (liver is the body's storehouse of nutrients). They were all horrified a few days later when their skin began drying up and flaking off.

Because of this effect of drying the skin by removing natural oils high doses of Vitamin A are sometimes used to combat acne.

One study found Vitamin A levels four times higher than normal among a group of eight girls suffering from anorexia (self starvation).

Like most nutrients that are toxic in excess, body size makes a difference. A rowing crew once took large doses of Vitamin A, but the only one who developed typical overdose symptoms was the small coxswain who sits in the back of the boat calling the strokes.

Foods high in Vitamin A are often yellow or orange in colour, ie, carrots, pumpkin, apricots. Because of this another symptom of excess Vitamin A can be a yellow hue to our skin, especially the palms of our hands. Chickens fed a diet high in Vitamin A have yellow skin.

Role in our body
- Helps maintain clarity and lubrication of our eyes.
- Assists our eyes in adapting to darkness.
- Anti-oxidant, to maintain the youthfulness of body cells and assist our immune system kill infections and cancerous cells.
- Helps maintain the lubrication of internal linings and outer skin.
- Necessary for normal development and strength of bones.
- Helps regulate spacing and straightness of teeth in children.
- Protects against tooth decay.

Too little
- Irritable, inflamed, dry, dull, or rough-feeling eyes. Blindness in severe cases. Cataracts. Reduced night vision.
- Tuberculosis.
- Weak bones and teeth.
- Stunted growth and crooked teeth in children.
- Low immunity to infection.
- Cancers of body organs.
- Oily or coarse skin on face and back, sometimes with acne.
- Dull hair. Excessive dandruff.
- Reduced sense of smell.
- Rapid loss of Vitamin C.

Too much (over 7000 mcg daily)
- Itchy, dry, sometimes flaking skin, with raised hair follicles.
- Yellow tinge to skin, especially on palms.
- Excessive hair loss. Brittle nails.
- Loss of appetite, nausea and headaches. Blurred vision.
- Restlessness and insomnia.
- Birth defects if excess taken during pregnancy.

Cooking losses
No significant loss.

"I believe that modern nutritional deficiencies may account for over half of all disease."
                    Dr Donald Rudin. Author *"The Omega 3 Phenomenon."*

Vitamin A

	RDI	
Men	750 mcg	
Women	750 mcg	Breast feeding 1200 mcg

	DAIRY / EGGS	mcg		**VEGETABLES (cooked)**	
3 tbsp	Butter	450	½ cup	Silverbeet	2500
3 tbsp	Cream	200	¾ cup	Pumpkin	2000
2 cups	Milk (std)	170	½ cup	Carrots	1050
2 med	Eggs	140	2 cobs	Sweet Corn	405
3 tbsp	Margarine	100	½ cup	Spinach	170
½ cup	Ice Cream	75	1 cup	Broccoli	165
	FRUITS		1 med	Tomato (raw)	120
1	Apricot	465		**FISH**	
1 cup	Cantaloupe	270	100 gms	Eel	1900
5	Apricots (dried halves)	240	100 gms	Orange Roughy	100
1	Persimmon	180			
1 cup	Pawpaw	170		**HEALTH FOODS**	
2	Plums	140	100 gms	Liver	13,000
¾ cup	Peaches (canned)	120	1 tbsp	Cod Liver Oil	3600
1	Tamarillo	115	100 gms	Kidney	320
1	Peach	110			
½	Avocado	105			
1	Nectarine	85			

HEALTH HINT **Change your diet gradually**

Major changes in our diet should be phased in gradually to allow our digestive system time to develop the necessary enzymes to properly digest the new food, otherwise temporary nutrition deficiencies can occur.

For similar reasons we should when possible eat the same types of food at the same time each day. Our bodies thrive on regularity.

"To rest is to rust."   Hippocrates

"Homo Sapiens is the only animal that hasn't enough sense to stop eating when he is sick."
                                    Dr Ulric Williams. NZ surgeon turned natural healer.

Vitamin B1 – Thiamine

Thiamine is the first of the B vitamins we look at. All of the B family work together and are usually found together in food.

As you will see in the sections ahead, B group vitamins can have a marked effect on our emotional and mental states. Lack of Thiamine for example can bring on depression, apathy and confusion, along with feelings of impending doom. If that were not enough, severe lack can cause panic attacks, rapid heart beat and eventual death.

Thiamine is one of the most commonly deficient vitamins in the world today, especially among rice eating peoples who use white rice (which is quicker to cook) instead of the more nutritious unrefined brown rice. By removing the outer bran during refining, 70% of the Thiamine in rice is lost. Wheat also loses 60% of Thiamine when refined into white flour.

Surveys have found widespread Thiamine deficiencies among females and alcoholics. Over 50% of a group of pregnant women were found deficient in an Australian survey. NZ dietary intake appears satisfactory, but smoking, alcohol and excess sugar can deplete body reserves.

Most of the B group vitamins like Thiamine are harmless in excess. Being water soluble they can usually be flushed out of the body in the urine. However 95% of 230 cot death babies in one American study were found to have excess Thiamine in their blood, possibly indicating lack of absorption. Injecting excess Thiamine into animals causes death through breathing failure which is significant feature of cot deaths amongst infants.

There was also an Australian experience reported many years ago where an Aboriginal community, severely deficient in Thiamine were experiencing a cot death rate of 40% among breast-fed babies. This was able to be immediately rectified by supplementing both the mothers and the infants with the vitamin.

Travellers in lands infested with biting bugs sometimes take large doses of Thiamine (5000 mcg daily) to discourage bites.

Role in our body
- Promotes the calmness of our brain and nerves and a sense of well-being.
- Assists in the manufacture and regulation of fats and acids.
- Helps transport blood sugar and oxygen to needed areas.
- Helps maintain health of our digestive system and heart.
- Helps regulate the growth of children.

Too little
- Depression. Nervousness. Panic attacks. Insomnia.
- Irregular or rapid heart beat (palpitations).
- Apathy. Confusion. Poor memory. Mental disorders.
- Low energy. Loss of appetite. Weight loss. Constipation.
- Anorexia (factor).
- Attention disorders and/or slow growth and learning in children.
- Numbness and tingling in hands and feet.
- Prickly swollen, painful muscles in lower legs. Eventual loss of feeling in legs.
- Increased risk of cot death among infants.

Too much
No known toxic effects in adults, but excess has been found in the blood of cot death infants.

Cooking losses
Boiling food in water can leach out about a third of the Thiamine. This can be recovered if the water is consumed. About 70% of Thiamine in peanuts is lost when they are roasted.

"A person's mental health can be measured by the amount of good he or she sees in the people around them."
Traditional maxim

Vitamin B1 Thiamine

	RDI	
Men	1100 mcg	
Women	700 mcg	

DAIRY / EGGS

		mcg
2 cups	Milo (all std milk)	860
2 cups	Milk (std)	360
2 cups	Milk (low fat)	160

FRUITS

1 cup	Orange Juice (pure)	180

GRAINS / NUTS

1/3 cup	Brazil nuts/Peanuts (raw)	450
1 plate	Cornflakes (with milk)	390
½ cup	Flour - Wholemeal	260
4 slices	Bread - Wholemeal	240
½ cup	Flour - White	150
1/3 cup	Other Nuts (avg)	125
1/3 cup	Peanuts (roasted)	125
1 plate	Porridge (with milk)	120

MEATS / FISH (cooked)

100 gms	Pork	750
100 gms	Trout (Brown)	320
100 gms	Lamb	130
100 gms	Beef	120

VEGETABLES (cooked)

2 cobs	Sweet Corn	520
¾ cup	Peas	305
2 med	Potatoes	150
¾ cup	Baked Beans	140
1 med	Kumara	140
1 cup	Broccoli	100
1 cup	Brussels Sprouts	100
1 cup	Cauliflower	100

HEALTH FOODS

1 tbsp	Brewers Yeast	1250
1 tsp	Marmite/Vegemite	550
100 gms	Kidney (Lamb)	480
1 tbsp	Malt extract	480
1/3 cup	Oat Bran	450
1 plate	Muesli (with milk)	380
3 tbsp	Sesame Seeds	300
100 gms	Liver (Lamb)	260
100 gms	Kidney (Beef)	250
2 tbsp	Wheatgerm	200
100 gms	Liver (Beef)	180

HEALTH HINT **Negative thinking and health**

A major root cause of mental and physical ill health is the habit of negative thinking. Lack of B group vitamins such as Thiamine can contribute to habitual negativeness, which often leads to depression and other mental and physical disorders.

Psychologists say that a simple way to break this destructive habit is to remember every morning on awakening, a happy event that has occurred in our past life. By holding this happy memory in the forefront of our mind throughout the day we can largely prevent negative thoughts from gaining a foothold.

Vitamin B2 – Riboflavin

Riboflavin is a fluorescent green-yellow vitamin in appearance and is sometimes used as a food colouring (code 101). It is also this quality that can cause our urine to turn a bright yellow after taking a multi-vitamin capsule. See also comment on page 143.

Riboflavin was found to be a commonly deficient vitamin in the 1997 NZ nutrition survey, especially among teenage girls and Maori women over 45. American surveys have also found deficiencies among pre-school children (not included in the NZ survey).

An active lifestyle increases our requirement for this vitamin and research indicates also that women taking the birth control pill need higher levels of Riboflavin in their diet, along with Vitamins B6, B12 and Folate.

Sunlight destroys Riboflavin – 70% is lost from milk after 3½ hours in direct sunlight. Baking soda can also destroy this vitamin.

Riboflavin like Vitamin A is important for healthy eyesight. Overseas research has linked long term Riboflavin deficiency with cataracts of the eyes. Many elderly people give up drinking milk which is their main source of Riboflavin. Continued treatment with Riboflavin has been reported to cure cataracts if caught early.

Role in our body
- Helps us absorb nutrients from food.
- Helps regulate the growth of children.
- Helps maintain health of our eyes.
- Assists the transport of blood sugar to muscles for energy.
- Helps our cells obtain oxygen from the blood.

Too little
- Irritable, bloodshot eyes. Over-sensitivity to bright light.
- Dark skin under eyes. Red eyelids.
- Eye cataracts (loss of transparency of lens).
- Slow growth in children.
- Insufficient milk in breast feeding mothers.
- Red greasy skin on face.
- Cracks on lips or at the corners of mouth.

- Red burning tongue or lips.
- Eczema or scaly skin on face and/or genitals. Itching skin.
- Excessive hair loss.

Too much
No known toxic effects.

Cooking losses
No significant losses, however baking soda which is an ingredient of baking powder can destroy Riboflavin.

HEALTH HINT   Milk a good source of Riboflavin

Milk is a rich natural source of Riboflavin. Where budgets are limited, non-fat milk powder can be bought for about half the cost of the equivalent liquid milk. By mixing it a little richer than recommended, a pleasant drink can be obtained that is higher in Riboflavin than regular milk.

Are NZ foods contaminated by pesticides?

As part of the 1997 nutrition survey, 114 NZ foods were tested for contamination by 90 different pesticides. Residues of 20 pesticides were found in 59% of the foods but in all cases they were well below the TWI (tolerable weekly intake).

Breathing spray drift of pesticides would therefore appear to present a greater danger than ingesting them in food.

As regards other contaminants, only Cadmium was of concern, due to past contamination in imported Nauru fertiliser and naturally high Cadmium levels of Bluff oysters.

Lead levels in food had dropped by more than half since the last survey and were well within normal safety limits.

Vitamin B2 Riboflavin

	RDI		
Men	1400 mcg		
Women	1000 mcg	Pregnancy 1300 mcg	Breast feeding 1500 mcg

		mcg			
	DAIRY / EGGS		100 gms	Chicken	150
2 cups	Milo (all milk)	1240	100 gms	Beef	150
2 cups	Milk (std or non-fat)	1000	100 gms	Fish (avg)	120
2 med	Eggs	300		**VEGETABLES (cooked)**	
½ cup	Ice Cream	240	1 cup	Broccoli	330
	NUTS		2 cobs	Sweet Corn	200
1/3 cup	Almonds	470	¾ cup	Peas	140
	MEATS / FISH (cooked)			**HEALTH FOODS**	
100 gms	Lamb	500	100 gms	Liver	3600
100 gms	Venison	410	100 gms	Kidney	2100
100 gms	Eel	400	1 plate	Muesli (with milk)	700
100 gms	Mackerel	380	1 cup	Soy Milk (full fat)	480
6	Oysters	360	1 tbsp	Malt extract	450
100 gms	Sardines (canned)	360	1 tsp	Marmite/Vegemite	430
100 gms	Brown Trout	320	1 tbsp	Brewers Yeast	340
100 gms	Pork	250			
100 gms	Salmon (canned)	180			

HEALTH HINT   Eggs a rich source of nutrients

Eggs are a rich source of readily available nutrients. This is not surprising when we consider that a fully formed live chicken, complete with eyes, down, beak, bones and all other body parts can be formed from the nutrients of just one egg.

"I am come down to deliver them out of the hand of the Egyptians, unto a good land, a land flowing with milk and honey."
God speaking in the Bible book of Exodus 3:8.

Vitamin B3 – Niacin

Niacin is sometimes called Nicotinic Acid but is not related to the nicotine found in tobacco.

Lack of Niacin can have a marked effect on our emotional state of mind and result in a constant state of needless anxiety. This can often lead to tranquilliser, tobacco, or alcohol dependence.

60% of alcoholics being treated in an American hospital study benefited by a rectification of their Niacin deficiency, which is almost universal among alcoholics. They reported a reduced craving for alcohol, and 20% maintained complete sobriety after discharge compared to 0% under previous treatment.

Niacin deficiency has also been linked with arthritis and high blood cholesterol levels. High supplements of over 100 mg daily have been found to lower harmful low density cholesterol levels 30%, and increase good high density cholesterol 35%.

Regular very high supplements of up to 3000 mg daily have been found to cure many schizophrenics who often have higher than normal needs, and also those with other mental disorders. Niacin can take the place of tranquillisers.

100 mg of Niacin daily for two weeks can be an effective treatment for acne. All high doses of Niacin should be in the form known as Niacinamide to avoid side effects.

Widespread deficiencies of Niacin are found in areas of the world where the diet is predominantly corn. NZ intake is generally satisfactory.

Like Riboflavin, an active lifestyle increases our requirement.

Role in our body
- Helps us absorb nutrients from food.
- Calms and regulates the brain.
- Helps regulate cholesterol levels.
- Helps regulate blood sugar levels and provide energy.
- Helps maintain health of our nerves, skin, digestive and reproductive systems.

Too little
- Anxiety. Insomnia. Mental disorders.
- Low energy.

- Alcohol or drug addiction. Trembling.
- Migraine headaches.
- High cholesterol levels. High blood pressure. Heart attack. Stroke.
- Loss of hearing.
- Low appetite. Anorexia.
- Red burning tongue. Gum disease.
- Darkened or red skin on face and hands. Red eyelids.
- Arthritis (factor).

Too much

No known toxic effects, however dosages over 100 mg can produce a 15 minute, itchy flush to the face and neck. This is avoided if a modified form known as Niacinamide is used.

Cooking losses

No significant losses.

HEALTH HINT — Cheerful people live 19% longer

If we cannot smile readily, chances are we are under emotional stress.

A twenty year study has found that people with a cheery optimistic personality live on average 19% longer than negative thinking pessimists.

This is in line with recent findings by medical researchers who have experimentally demonstrated that an optimistic outlook, even a simple smile, strengthens our immune system.

Vitamin B3 Niacin

	RDI					
Men	19.0 mg					
Women	11.0 mg	Pregnancy	13.0 mg	Breast feeding	16.0 mg	

BEVERAGES (mg)
2 cups	Milo/Ovaltine (all milk)	5.0
1 can	Beer	1.5

DAIRY / EGGS
2 cups	Milk	4.2
2 med	Eggs	3.8

FRUITS
1 cup	Honeydew	2.1
1 cup	Pawpaw	1.7
½	Avocado	1.5
1 cup	Cantaloupe	1.4

GRAINS / NUTS
1/3 cup	Peanuts (roasted)	9.3
1 plate	Cornflakes (with milk)	4.5
4 slices	Bread - Wholemeal	4.0
4 slices	Bread - White	3.2
1 plate	Wheatflake cereal	3.2
½ cup	Flour - Wholemeal	2.6
1/3 cup	Other nuts (avg)	2.5
1 tbsp	Peanut Butter	2.4
½ cup	Flour - White	2.1
½ cup	Brown Rice	2.0

MEAT / FISH (cooked)
100 gms	Chicken	7.0
100 gms	Lamb	4.5
100 gms	Beef	4.0
100 gms	Pork	3.0
100 gms	Fish	3.0

VEGETABLES (cooked)
2 cobs	Sweet Corn	5.4
¾ cup	Peas	2.8
¾ cup	Baked Beans	2.6
1 bag	Potato Crisps (50gm)	2.5
2 med	Potatoes	2.0
1 cup	Broccoli	2.0
1 med	Kumara	1.9

HEALTH FOODS
1 tbsp	Malt extract	25.0
100 gms	Liver (lamb)	20.0
100 gms	Liver (beef)	15.5
1 cup	Tomato Juice	6.4
1 plate	Muesli (with milk)	6.0
2 tbsp	Pumpkin Seeds	3.4
1 cup	Soy Milk	3.3
1 tbsp	Brewers Yeast	3.0
1 tsp	Marmite/Vegemite	2.5

"A man without a purpose is like a ship without a rudder."

Carlyle, 1880.

Vitamin B6

Vitamin B6, like other B vitamins can have a profound effect on our mind. Do you recall dreams easily and clearly? If not you are possibly not getting enough B6. Total lack of dream recall can be a symptom of B6 deficiency. Over-vivid dreams can result from too much.

Lack of B6 has also been linked with hyperactivity and ADD (Attention Deficit Disorder) in children, although food allergies are also recognised as a major cause of this problem.

Vitamin B6 is necessary for good digestion and in particular to allow our body to absorb the Vitamin B12 and the minerals Iron and Zinc. Often a low blood count (anaemia) is caused by lack of B6 in the diet rather than lack of Iron or Zinc.

Convulsions in infants and also cleft palates have been strongly linked to lack of B6 during pregnancy.

Extra B6 (over 50,000 mcg daily) has been found to relieve pre-menstrual tension (PMT or PMS) and fluid retention which causes body tissues to swell, and also nausea and toxaemia in pregnant women.

100,000 mcg a day for three months has been reported to heal RSI or Carpal Tunnel Syndrome.

Anorexia (self starving) in girls, and some cases of Parkinson's Disease (uncontrolled trembling) have also been reported cured by extra B6, usually by injection.

Extra B6 can reduce our tendency to sunburn and promote tanning. When applied as a cream it has also been known to regress Melanoma type skin cancer.

Excess coffee, the birth control pill, and continued stress can lower body levels of Vitamin B6.

A simple test for a deficiency of B6 is to try and touch with your fingertips, the palm of your hand at the base of your fingers without bending them at the knuckles, only at the bottom finger joints. Failure to do so indicates that you are likely to be deficient.

Role in our body
- Helps us absorb nutrients from food – especially Protein, Iron, Zinc and Vitamin B12.

- Helps to calm and regulate the brain.
- Assists in releasing stored blood sugar from the liver.
- Assists in forming blood cells.
- Helps regulate the salt balance of cells.

Too little
- Toxaemia in pregnant women.
- Convulsions or mouth malformations in infants if mother is deficient. Stillbirths in serious cases.
- Anxiety, depression, insomnia and irritability, especially in pre-menstrual women (PMT).
- Hyperactivity and/or slow learning in children (ADD).
- Weak red blood cells (anaemia). Pale skin. Iron or Zinc deficiency. Low energy. Trembling.
- Stiff and swollen finger joints, sometimes lumps form.
- Numb or tingling hands, arms and legs. RSI or Carpal Tunnel Syndrome (factor).
- Swelling of legs and other parts of the body (edema or water retention).
- Cracks or sores at the corners of mouth. Sore red tongue. Red eyelids.
- Acne around nose or on forehead.
- Lack of dream recall.
- Twitchy eyelids. Cataracts.
- Low appetite. Nausea. Anorexia.
- Arthritis.

Too much
No known toxic effects under 50,000 mcg, however excess B6 can produce over-vivid dreams. Supplements over 20,000 mcg daily can cause a dependency on the supplement, ie our body may fail to extract B6 from food when the supplement is stopped.

Cooking losses
No significant losses.

"A longer shelf life for foods means a shorter shelf life for humans."　　　　　　　　　　　　　　　　　Toni Jeffreys Ph.D.

Vitamin B6 Pyridoxine

RDI		
Teenage boys	1800 mcg	
Teenage girls	1500 mcg	
Men	1500 mcg	
Women	1000 mcg	Pregnancy and breast-feeding 1800 mcg

		mcg
	DAIRY / EGGS	
2 cups	Milk (low fat)	260
2 cups	Milk (std)	200
2 med	Eggs	200
	GRAINS / NUTS	
1 plate	Cornflakes (with milk)	550
½ cup	Flour - Wholemeal	250
½ cup	Rice Brown (cooked)	190
	MEAT / FISH Cooked)	
100 gms	Chicken	560
100 gms	Liver	500
100 gms	Salmon/Sardines	460
100 gms	Pork	420
100 gms	Beef	240
100 gms	Fish	200
	FRUIT	
1	Banana	650
1 cup	Honeydew	240
½	Avocado	210
1	Kiwifruit	160
1/3 cup	Raisins/Sultanas	150
	VEGETABLES Cooked)	
2 cobs	Sweet Corn	700
1 bag	Potato Crisps (50gm)	400
1 cup	Brussels Sprouts	280
¾ cup	Baked Beans	240
1 cup	Broccoli	210
½ cup	Spinach	200
1 cup	Cauliflower	180
1 med	Kumara	170
2 med	Potatoes	160
	HEALTH FOODS	
1 plate	Muesli (with milk)	1000
1 tbsp	Malt extract	750
1 cup	Tomato Juice	300
1 tbsp	Brewers Yeast	200
2 tbsp	Wheatgerm	200
1 cup	Soy Milk	150

HEALTH HINT — **Pregnancy after the Pill**

Gross metabolic disturbances occur in a woman's body when she takes the Pill and it would appear best to avoid doing so if she wishes to enjoy long term health.

Women who are on the Pill and plan to eventually conceive should use some other method of birth control at least three months before attempting conception.

The following nutrients should also be increased during that period: Thiamine, Riboflavin, B6, B12, Folate, Vitamin C, Zinc and Magnesium.

HEALTH HINT   Three rules to minimise stress

Lack of Vitamin B6 can result in us being less effective in handling stress. Long term stress will greatly lower our immunity to sickness and is a major factor in poor health.

Most sources of stress involve just three areas in life:

- Family and friends.
- Finances.
- Fellow workers.

There are three simple, proven rules to minimise stress in these areas. They are as follows:

- Family and friends – forgive those who have offended us, either face to face or by a letter.
- Finances – spend <u>less</u> than we earn and invest 10% of our income.
- Fellow workers – if we cannot wake up most mornings and look forward to our day's work or activity we are in the wrong occupation. We should change it for work or activity that we enjoy, even if our income is lower.

Excess B6 can produce over-vivid dreams...

Folate (Folic Acid)

This bright yellow B vitamin is the most commonly deficient vitamin in NZ, especially among infants and women (Maori women in particular).

Excess alcohol, smoking and the pill are known to hinder the absorption of Folate.

Folate is required for cell division and is especially important for the health of pregnant women and their babies. Their need doubles during pregnancy and deficiencies have been strongly linked with pregnancy problems such as toxaemia. These problems can result in miscarriage, mental retardation, Spina Bifida and malformities of the child, especially cleft palate.

Low levels are also found in heart attack victims and arthritics.

A recent large study, reported in 1999 found that adequate Folate <u>reduces the incidence of bowel cancer a massive 75%</u>.

So strong is the evidence of health problems linked with lack of Folate that American health authorities require white flour to be fortified with this vitamin, which has proved successful in reducing deficiencies.

Lack of Folate during times of severe anxiety has been linked with premature greying of hair, and interestingly a large amount of Folate can restore abnormally grey hair on an animal to its natural colour, but it does not seem to work with humans.

Folate is similar in action to B12 and appears to be equally important for mental health. Low levels are commonly found among alcoholics, mentally retarded children, psychiatric patients and those suffering from senility or Alzheimer's disease.

Many schizophrenics respond to a combined treatment of Folate, Niacin and Zinc, and epileptics have been found to improve on Folate and B12 supplements.

It is believed that our body can manufacture small amounts of Folate but most must be supplied from our food.

Role in our body

- Helps our body absorb and utilise protein.
- Assists in the forming of body and blood cells.
- Necessary for healthy development of a foetus during pregnancy.
- Assists in maintaining normal brain function.

- Assists in maintaining clear arteries.
- Helps maintain the health of our eyes, hair and skin.
- Helps maintain health of our digestive system.
- Helps control uric acid levels.

Too little
- Toxaemia, premature birth or miscarriage in pregnant women.
- Malformed and retarded infants if mother deficient.
- Irritability. Senility. Mental disorders. Alcohol and drug addiction. Alzheimer's disease.
- Weak red blood cells, pale skin, low energy (anaemia).
- Clogged arteries. Heart Attack.
- Cold hands and feet.
- Sore smooth red tongue. Stomach ulcers.
- Weak vision in elderly.
- Attention disorders in children.
- Premature grey hair.
- Gout. Arthritis.
- Lung cancer in heavy smokers (factor).

Too much (over 15,000 mcg daily)
- Nausea. Wind. Insomnia.

Cooking losses
High – usually around 50% if cooked at high temperatures.

The need for Folate doubles during pregnancy – oranges are a good source.

Folate

	RDI	
Men	200 mcg	
Women	200 mcg	Pregnancy and breast feeding 400 mcg

DAIRY / EGGS (mcg)
		mcg
2 med	Eggs	40
2 cups	Milk - (std or low-fat)	25

FRUITS
1 cup	Honeydew	100
½	Avocado	55
1	Orange	50
1 cup	Rock Melon	40
1 cup	Orange Juice (pure)	20

GRAINS / NUTS
1 plate	Cornflakes (with milk)	85
1/3 cup	Peanuts (roasted)	65
4 slices	Bread - Wholemeal	40
1/3 cup	Almond / Cashew nuts	40
½ cup	Flour - Wholemeal	35
4 slices	Bread - White	30
2	Weetbix (with milk)	20

MEAT / FISH Cooked
100 gms	Pipis	50

VEGETABLES (cooked)
1 cup	Broccoli	180
1 cup	Brussels Sprouts	145
½ cup	Spinach	120
2 cobs	Sweet Corn	85
¾ cup	Green Peas	75
1 cup	Cauliflower	70
4 slices	Beetroot	60
¾ cup	Baked Beans	50
1 med	Kumara	35
1 med	Tomato (raw)	30
½ cup	Green Beans	25
2 med	Potatoes	25
1	Onion	25
½ cup	Coleslaw	20
1 cup	Lettuce (raw)	20

HEALTH FOODS
100 gms	Liver	260
1 tbsp	Brewers Yeast	190
1 tsp	Marmite	100
1 plate	Muesli (with milk)	95
1	Muesli Bar	45
2 tbsp	Wheat Germ	40
1 cup	Soy Milk	25

HEALTH HINT — A healthy pregnancy

The need for Folate doubles during pregnancy. This is best achieved with natural foods rather than supplements. Mothers-to-be should therefore eat plenty of green vegetables, yellow melons and oranges throughout their pregnancy.

Also a long walk most days minimises the common risks of varicose veins and constipation.

"The best time to start feeding a baby is several years before it is born." — Brenda Sampson. NZ allergy consultant & author.

"As our body ages we require more nutrients and fewer calories."
Jane Kinderlehrer, American lecturer on nutrition.

Vitamin B12

Many disorders have been linked with lack of this very important vitamin.

B12 like the other B vitamins is a brain nutrient and necessary for a sense of well being and relaxation. However unlike the other B vitamins, B12 is not normally found in fruits, grains and vegetables, but in animal products such as milk, eggs and meat. It has also been found in natural water supplies. Liver and pipis are a rich source.

As might be expected, deficiencies have been found among strict vegetarians, but mostly in the western world. Deficiencies are uncommon among the people of poorer nations, many of whom live solely on a vegetarian diet. One likely explanation is that B12 which is only required in minute amounts, is obtained from untreated drinking and cooking water and from insect life, especially eggs consumed along with plants. Not all peoples tend to be as fanatical about food hygiene as Westerners.

The mineral Cobalt is part of B12. Some NZ soils are seriously low in this mineral and farm stock have suffered badly in the past. Cobalt is now added to fertiliser where necessary.

One American researcher found levels of Cobalt to be very low in violent offenders by using hair analysis.

Even when animal products are part of the diet, deficiencies of B12 are widespread among the elderly (especially the blue eyed), many of whom appear to have difficulty in absorbing B12 from food. A UK survey during 2001 found 50% of the elderly to be deficient. Monthly injections of the vitamin are often given to overcome this problem. Our liver can normally store over four years supply.

Often schizophrenia-like symptoms occurring after 40 years of age, such as hearing voices in the head, poor memory and mental confusion, are the result of B12 deficiency. These symptoms are often put down to senility in the elderly.

Other disorders strongly linked with lack of B12 are multiple sclerosis (a nerve disease), osteoporosis, arthritis, listlessness, food or chemical allergies, and patchy loss of pigmentation in brown skinned people.

Excess sugar consumption and the birth control pill are also known to lower body levels of B12.

Role in our body
- Helps us absorb nutrients from food.
- Assists in the forming of red blood cells.
- Helps maintain health of bones and bone marrow.
- Helps regulate brain activity.
- Helps maintain health of nerve fibres.
- Assists in regulating fat storage.

Too little
- Schizophrenia and senility type symptoms. Insomnia.
- Weak red blood cells (anaemia). Pale skin. Low energy.
- Painful joints, sometimes with brown discolouration.
- Calcium growths. Osteoporosis and Arthritis (factor).
- Reduced sense perception (weak eyesight, hearing, touch, etc). Poor balance.
- Weak, painful, poorly controlled muscles, difficulty in walking smoothly, especially in the dark.
- Numb, tingly, legs and fingers.
- Multiple sclerosis (factor).
- Easy bruising.
- Dark red tongue. Swollen face tissue.
- Premature grey hair.
- Patchy loss of skin pigmentation in brown skinned people.
- Lung cancer in heavy smokers (factor).
- Food and chemical allergies.

Too much (Cobalt poisoning)
- Iodine deficiency. Swollen thyroid gland (Goitre).
- High cholesterol levels. High blood pressure. Heart attack. Stroke.

Cooking losses
No significant losses

"A merry heart doeth good like a medicine."
King Solomon

Vitamin B12

	RDI		
Men	2.0 mcg		
Women	2.0 mcg	Pregnancy	3.0 mcg

	DAIRY / EGGS	mcg		**HEALTH FOODS**	
2 med	Eggs	2.0	100 gms	Liver (beef)	110.0
2 cups	Milk (std or low-fat)	1.8	100 gms	Liver (lamb)	81.0
25 gms	Cheese	.3	100 gms	Kidney (lamb)	79.0
½ cup	Ice cream	.2	100 gms	Kidney (beef)	31.0
			1 tsp	Marmite	.5
	MEAT / FISH (cooked)				
100 gms	Pipis	62.0			
100 gms	Sardines (canned)	28.0			
6	Oysters	18.0			
100 gms	Salmon (canned)	4.0			
100 gms	Lamb	2.2			
100 gms	Beef	1.6			
100 gms	Chicken	.8			
100 gms	Pork	.7			
100 gms	Fish	.6			

Signs of Alzheimer's disease

The following are generally regarded as reliable signs of Alzheimer's disease but lack of B12 can mimic these signs and should be suspected in those under 75.

1. Inability to count backward from 20, or to draw a clock face.
2. Unawareness of year or day of week.
3. No longer able to operate normal household appliances.
4. Getting lost in familiar surroundings.

HEALTH HINT   Milk can supply B12 needs

If strict vegetarianism appeals, the minor compromise of two glasses of milk daily (whole or low-fat) should take care of Vitamin B12 needs.

Lecithin

Lecithin is a waxy, fatty type substance, usually yellow in colour. It is found in egg yolk, animal fats and plant oils such as Soya bean oil, which is the main commercial source. It can also be manufactured by a healthy liver.

Lecithin is widely used as a food additive especially in margarine and ice cream to maintain a smooth consistency.

Being a fat it is high in calories, a mere 10 gms of Lecithin contains 90 calories.

Lecithin is widely believed to protect against heart disease by maintaining the optimum level of useful, high-density blood cholesterol. It is also believed essential for the efficient operation of our short-term memory.

Role in our body
- Helps maintain health of our body cells and brain.
- Helps maintain a healthy nervous system.
- Helps maintain levels of useful cholesterol.

Too little
- Impaired short-term memory.
- Nervous twitching.
- Maniac-depressive mental disorders.
- Increased risk of heart disease.
- Increased risk of hepatitis.

Too much
- Depression.
- Fishy body odour.

RDI
Not set as Lecithin is found in all fatty foods and can be manufactured by a healthy liver.

NO FOOD TABLE

Inositol

A lesser known B Group vitamin that is sometimes given as a supplement to treat high blood pressure, high cholesterol and anxiety. Symptoms that are often interrelated.

Many patients have thrown away their tranquillisers after a course of Inositol treatment – usually 1000 grams a day.

Other proven benefits of adequate Inositol are improved sleep and better absorption of Zinc.

Inositol can be manufactured in small amounts by our liver but excessive caffeine can deplete body reserves.

Role in our body
- Promotes calmness of the brain and a sense of well-being.
- Helps us absorb nutrients from food.
- Assists body to digest fats by manufacturing Lecithin.
- Helps maintain health of bone marrow, hair, eyes and intestines.
- Helps maintain health of our arteries, liver, kidneys and heart.

Too little
- Anxiety. Insomnia. Mental disorders.

Too much
No known toxic effects.

Cooking losses
No significant losses.

HEALTH HINT — **Three ways to a good sleep**

There is probably no greater aid to our mental and physical health than the rejuvenating effects of a good night's sleep.

If events of the day have made us feel tense before bedtime, here are three traditional ways to bring on the relaxed state essential for sound sleep.
- A glass of warm milk (unsweetened).
- A warm bath.
- Reading wholesome, but mildly boring literature.

Inositol (limited data available)

RDI 500 mg
Can be manufactured in small amounts by a healthy liver.

		mg
	DAIRY / EGGS	
2 cups	Milk - (std or low-fat)	100
	FRUITS	
1 cup	Orange Juice (pure)	340
1	Orange	250
1 cup	Cantaloupe	250
	GRAINS / NUTS	
1/3 cup	Peanuts (roasted)	90
4 slices	Bread - Wholemeal	90
½ cup	Rice - Brown (cooked)	80
4 slices	Bread - White	60
½ cup	Flour - Wholemeal	60
1 plate	Porridge	60
	MEATS / FISH (cooked)	
100 gms	Beef	150
100 gms	Chicken	50
100 gms	Pork	50
100 gms	Lamb	50
	VEGETABLES (cooked)	
¾ cup	Baked Beans	320
¾ cup	Peas	190
1 cup	Broccoli	125
1 cup	Brussels Sprouts	120
2 cobs	Sweet Corn	120
1 med	Kumara	100
½ cup	Spinach	90
½ cup	Silverbeet	85
1 med	Tomato (raw)	75
1 cup	Cabbage	75
1 cup	Lettuce (raw)	60
2 med	Potatoes	60
	HEALTH FOODS	
100 gms	Liver (cooked)	340
2 tbsp	Wheatgerm	120

Inositol has a natural, tranquillising effect.

Pantothenic Acid

Pantothenic Acid is a B group vitamin found in small amounts in nearly all foods. It can also be manufactured by a healthy body.

This vitamin is important to a healthy digestive system and is sometimes prescribed by surgeons to get the stomach working again after an operation.

Severe deficiencies are rare but low levels are often found in those who suffer from rheumatoid arthritis. In one trial involving twenty patients, arthritic pain was reduced markedly by daily injections of 50mg of Pantothenic Acid.

Pantothenic Acid like the other B vitamins is believed to help us handle emotional stress with greater composure. Emotional stress appears to be a key factor in arthritis.

A deficiency in animals is known to cause grey hair and restoring the deficiency restores the original colour, but as with Folate, experiments with large doses have failed to restore human hair to its former hue.

Better results may be achieved with longevity – Pantothenic Acid at three times normal intake increases the average life span of mice by 20%, to the human equivalent of 90 years.

Role in our body
- Necessary for the working of our digestive system.
- Assists our adrenal glands to produce hormones.
- Assists our immune system to form antibodies.
- Helps protect our body cells from the effects of radiation.
- Assists in the manufacture of body fats.
- Helps maintain the health of the brain, nervous system and skin.

Too little
- Loss of appetite. Indigestion. Low energy.
- Depression. Insomnia.
- Lowered resistance to radiation.
- Low immunity to colds and other infections.
- Burning sensation in feet.
- Cramp in arms or legs.
- Arthritis (factor).

Too much
No known toxic effects.

Cooking losses
About one third is destroyed if food cooked at high temperatures. Baking soda can also destroy Pantothenic Acid.

HEALTH HINT — Preventing acid stomach

The feeling of an acid stomach (which strangely can sometimes be due to lack of acid) is mostly the result of eating too much food, too quickly in a state of anxiety. Stomach acid is ejected past the upper valve or lid of the stomach and burns the esophagus tube. A tight belt can also contribute to this.

If we must eat when anxious (fasting has a calming effect on our mind), small, easily digested vegetable meals are preferable.

Long standing stomach irritation and excess wind can mean that digestive enzymes are lacking. These can be supplemented.

HEALTH HINT — Work off that stress

The best antidote to stress is vigorous exercise in the open air.

"Always rise from the table with an appetite and you will never sit down without one." — William Penn.

Pantothenic Acid

RDI 8.0 mg
Can be manufactured by a healthy body

	DAIRY / EGGS	mg		**VEGETABLES (cooked)**	
2 cups	Milk (low fat)	1.9	6 med	Mushrooms	1.4
2 cups	Milk (std)	1.5	1 cup	Pumpkin	1.2
2 med	Eggs	1.2	¾ cup	Split Peas	1.0
			2 cobs	Sweet Corn	1.0
	FRUITS		1 cup	Broccoli	.8
½	Avocado	.9	1 cup	Cauliflower	.8
			1 med	Kumara	.8
	GRAINS / NUTS		1 cup	Honeydew	.8
1/3 cup	Peanuts (cooked)	.9	2	Potatoes	.8
4 slices	Wholemeal Bread	.8			
				HEALTH FOODS	
	MEATS (cooked)		100 gms	Liver (lamb)	7.6
100 gms	Chicken	1.4	100 gms	Liver (beef)	5.7
			100 gms	Kidney (beef)	3.0
			100 gms	Kidney (lamb)	2.0

HEALTH HINT — The power of our words

B Vitamins are not the only things that can markedly affect our sense of well being. Whether we feel positive relaxed and healthy, or negative and tense is also largely determined by the words we choose to speak.

How much better we and all the other people around us would feel if the only words we ever spoke first passed the test of being – **Kind, True,** and **Necessary**.

"God grant me the serenity to accept things I cannot change, the courage to change the things I can, and the wisdom to know the difference."
Anonymous

Vitamin C

Up until 200 years ago it was not unknown for two thirds of a ship's crew to die of scurvy during a long sea voyage. In 1577 a Spanish galleon was found drifting with all on board dead from scurvy.

This unpleasant disease is caused by lack of Vitamin C and was due to the absence of fruit and vegetables on board. The English navy solved the problem with lime juice, making possible long voyages of discovery by explorers such as James Cook.

Vitamin C could well be described as an anti-toxin vitamin. A toxin is any substance harmful to the body. Vitamin C assists our body in ridding itself of toxic metals such as Cadmium, Lead and Mercury, and in some individuals, especially arthritics, excess Iron.

Vitamin C is also an important anti-oxidant and helps protect our body cells from the aging effects of oxidation.

Another quality of Vitamin C is the ability to lower blood cholesterol levels. 1000 mg daily usually brings about a 10% reduction in low density cholesterol and 40% of other blood fats, with a corresponding drop in blood pressure.

1000 mgs of Vitamin C daily has also been reported to help prevent and speed recovery of naturally occurring colds, 37% faster in one study. It was not however as effective with artificially induced colds.

In another study, 200 mgs of Vitamin C daily brought about a 15% reduction in natural colds, and total avoidance of colds was achieved at 3000 mg daily, provided the dosage was temporarily increased to 4000 mg at the first symptom of a cold.

Increased mental alertness is reported among those with high intakes of this vitamin. Kiwifruit and oranges are good sources.

Vitamin C is found in human milk in useful amounts, but is lacking in cow's milk. Unlike humans, animals can manufacture their own Vitamin C. Bottle-fed babies therefore need to have their milk supplemented. This is usually done with fruit juice.

The 1997 nutrition survey found that the average Vitamin C intake in NZ was a surprisingly high 114 mgs daily (RDI 35 mg). However widespread deficiencies have shown up in overseas surveys, especially among asthmatics, elderly men, 75% of cancer patients, and up to 100% of schizophrenics and other psychiatric patients.

One USA study showed an average of six days for psychiatric patients to reach tissue saturation of Vitamin C compared to only one and a half days for a control group of normal people. It was also noticed that the patients receiving the high doses smiled more often, were less tense and more sociable. High dosages have also been successful in assisting drug addicts overcome their addiction.

An English experiment involving 283 pregnant women with a history of pre-eclampsia (high blood pressure, kidney damage and fits) cut the incidence by 75% with high doses of Vitamin C and E.

Smoking destroys on average 27% of the Vitamin C in the blood each day, which contributes to the haggard appearance of many smokers. Aspirin has a similar effect.

Role in our body
- Helps us absorb nutrients from our food, especially Iron.
- Assists our immune system to kill infections, and to eliminate toxic metals and poisons.
- Assists in forming and maintaining the health of our bones and teeth.
- Protects cells from oxygen damage (anti-oxidant effect).
- Assists in forming body and blood cells.
- Helps heal broken bones, bruises and wounds.
- Helps regulate our cholesterol levels.

Too little
- Depression. Anxiety. Nervousness. Insomnia. Mental disorders. Addictions.
- High cholesterol levels. High blood pressure. Heart attack. Stroke.
- Pre-eclampsia during pregnancy. Kidney disorders.
- Increased levels of toxins in body.
- Low immunity to colds, infections and poisons, and slower recovery. Bed sores.
- Slow wound and bone fracture healing.
- Eye cataracts. Glaucoma.
- Abnormally red gums.

- Easy bruising. Purplish skin.
- Painful joints and lower spine, internal bleeding (Scurvy).
- Bleeding gums. Loose, easily decayed teeth.
- Haggard face. Dry skin. Scaly skin. Raised hair follicles. Premature aging. Pin head spots.
- Weak red blood cells, low energy, pale skin (Iron deficient anaemia).
- Insufficient milk in breastfeeding mothers.
- Food and chemical allergies.
- Throat, stomach and cervical cancers.
- Male infertility.
- Poor digestion of food. Bad breath.
- Arthritis and muscle pain.
- Asthma (factor).

Too much
Over 2000 mg daily can cause wind, diarrhoea, skin rashes or irritation in urinary passage.

Cooking losses
Boiling food in water can wash out much of the Vitamin C but this can be recovered if the water is consumed. Long exposure of food to air can destroy Vitamin C.

HEALTH HINT   **An apple a day . . .**

New research has discovered that eating just one apple a day, with the peel still on, can provide as much anti-oxidant activity as 1500 mg of Vitamin C taken as a supplement. Evidently the phenolic acids found in apple skin greatly enhance anti-oxidant activity.

"Let food be your medicine and medicine be your food."
                                                Hippocrates

Vitamin C

	RDI					
Men	40 mg					
Women	30 mg	Pregnancy	60 mg	Breast feeding	75 mg	

BEVERAGES

		mg
1 can	Beer	7

DAIRY / EGGS

		mg
2 cups	Milk (std or low-fat)	7

FRUITS

		mg
1	Kiwifruit	95
1 cup	Orange Juice	90
1	Tomato	90
1 cup	Honeydew	85
1 cup	Grapefruit Juice	70
1	Orange	65
1 cup	Rockmelon/Cantaloupe	45
1 cup	Apple Juice	30
1 cup	Pawpaw	21
1	Tamarillo	20
1 med	Banana	13
1/8	Watermelon	11
1 med	Apple	10
1	Persimmon	10
½ cup	Pineapple (canned)	9
2	Passionfruit	8
2	Apricots	8
1	Peach	8
1	Nectarine	6
½	Avocado	6
½ cup	Peaches (canned)	5
2 med	Plums	3
½ cup	Grapes	3

VEGETABLES (cooked)

		mg
½ cup	Green Peppers	85
1 cup	Brussels Sprouts	65
1 cup	Cauliflower	60
1 cup	Broccoli	55
1 med	Kumara	35
½ cup	Coleslaw	30
1	Tomato (raw)	30
½ cup	Cabbage	30
1 bag	Potato Crisps (50gm)	25
2 cobs	Sweet Corn	24
2 med	Potatoes	20
¾ cup	Peas	20
½ cup	Spinach	15
¾ cup	Pumpkin	15
¾ cup	Parsnips	12
½ cup	Alfalfa Sprouts (raw)	8
½ cup	Mung Bean Sprouts	7
1 cup	Lettuce (raw	7
1	Onion	7
½ cup	Green Beans	5
½ cup	Silverbeet	5
3 slices	Beetroot	5
½ cup	Carrots	3

HEALTH FOODS

		mg
1 tsp	Ascorbic Acid Powder	4000
1 cup	Lemon Juice	100
1 tbsp	Rosehip Syrup	60

HEALTH HINT — Vitamin C a healer

If you suffer from any of the following disorders you may wish to try taking 3000 mg of Vitamin C daily and see if any improvement occurs. All are on record as having responded to high doses of Vitamin C.

Infections including colds, high blood pressure, high cholesterol, arthritis (all types), gum disease, low energy, anxiety, insomnia, mental disorders and addictions.

Vitamin D

Vitamin D is a white crystalline substance, like salt, that is formed in our bodies by the action of sunlight falling on our skin. It can be stored in our liver for long periods.

The fairer our skin the more efficiently Vitamin D is formed, but it will not take place indoors as the necessary ultra-violet part of sunlight does not pass through window glass.

Because of high sunlight levels in New Zealand, deficiencies were long thought to be rare, except in babies who were bottle-fed on unsupplemented cow's milk. However a two year study in New Zealand found low levels of Vitamin D among heart attack patients, and also noted that their death rate was nearly 50% higher during the winter months. A follow-up Australian study came up with similar findings.

A more recent study also found that average Vitamin D levels have typically dropped 75% by age 70, indicating perhaps a lessening efficiency in our skin's ability to manufacture Vitamin D as we age. Although sleeping pills, cholesterol lowering drugs and cortisone can also block this action.

A nursing home experiment involving 3270 women, with an average age of 84, reduced the number of hip fractures by 43% with daily supplements of 20 mcg of Vitamin D and 1200 mg of Calcium.

Vitamin D is regularly added to milk overseas, especially in countries that do not see a lot Of sunlight.

Useful amounts are found in some oily foods especially fish oil. Cod Liver Oil is an especially rich source.

Vitamin D is necessary for the correct formation of bones and teeth in children. Calcium and other bone-forming minerals cannot be absorbed well from food without this vitamin, although some researchers are now claiming that Boron also plays a role.

Malformed bones, especially bowed legs can result from long term Vitamin D deficiency.

Role in our body
- Helps us absorb nutrients from food, especially Calcium.
- Forms and maintains the health of our bones and teeth.
- Enhances our immune system.

Too little
- Osteoporosis.
- Heart attack.
- Eye cataracts in infants.
- Weak bones in children, leading to bowed legs and spinal malformities (rickets).
- Poor muscle development in adolescents.
- Soft painful bones in adults, leg aches, weak muscles and tooth decay.
- Increased risk of tuberculosis.

Too much (by supplement only)
- Calcium build up in arteries, heart and kidneys.
- Kidney stones.
- High blood pressure.
- Loss of appetite. Nausea. Drowsiness. Frequent urination.

Cooking losses
No significant losses.

"But I must get my Vitamin D dear."

Vitamin D (limited data available)

RDI 10 mcg
Can be manufactured by the action of sunlight on skin.

		mcg	
3 hours	Sunshine (face only)	10	Mid-Summer
3 hours	Sunshine (face only)	2	Mid-Winter
	FOOD SOURCES		
1 tbsp	Cod Liver Oil	360	
100 gms	Fish - canned varieties	12	
3 tbsp	Butter/Margarine	5	
2 med	Eggs	2	

HEALTH HINT — Banish bad breath

Most moderate bad breath comes from stagnant bacteria in our mouth especially on the tongue. Eating will usually freshen this bacteria and give temporary relief.

A longer term solution is, morning and night when brushing our teeth to also vigorously brush our tongue with diluted toothpaste as far back as we can reach without gagging, also our gums, roof of mouth and other surfaces which might harbour bacteria.

Severe continual bad breath is usually caused by rotting teeth or tonsil infection and requires professional help.

HEALTH HINT — Brisk walking lowers cancer risk

A recent large study came to the conclusion that an hour a day of <u>brisk</u> walking cuts the risk of organ cancer 20%.

"Without health, riches, possession and fame are all mud."
Ed Wynn - Comedian

"There is at bottom only one genuine scientific treatment for all diseases and that is to stimulate the phagocytes (immune system). Drugs are a delusion."

George Bernard Shaw. 1856 -1950.

Vitamin E

Vitamin E is a powerful anti-oxidant that works hand in hand with Selenium to protect our cells from the aging effects of oxygen damage known as oxidation. Oxygen damage can be seen in the browning effect when a sliced apple is left exposed to the air.

Vitamin E therefore plays a vital role in our immune system along with the other anti-oxidants, Vitamin A, CoQ10, Vitamin C, Zinc and Selenium. These nutrients also help in ridding our body of toxic substances.

As well as being an anti-oxidant Vitamin E helps maintain the health and correct viscosity of our blood. Low levels of Vitamin E are consistently found in heart attack victims.

A 1996 study of 35,000 women aged over 50, found those whose diets had the most Vitamin E, had a 62% less chance of dying from heart disease compared to those whose diets had the least, but only if they got their Vitamin E from natural foods. The most active form of Vitamin E is only found in plant life. Women who took the vitamin in synthetic form continued to suffer heart disease at the average rate.

A double blind Finnish study of 27,000 older male smokers achieved a 32% reduction in Prostate cancer with just 5 mg extra Vitamin E daily.

Vitamin E has also been found to relieve cramps, sore breasts and leg swelling suffered by many women prior to menstruation.

Vitamin E intake in NZ appears to be generally adequate. But if we consume much cooked unsaturated fat then it is important that we obtain the full RDI amount of Vitamin E to protect our cells from the harmful oxidation effect of free radicals which form in unsaturated fats when they are heated. Fortunately the best sources of Vitamin E are found in unsaturated fats, especially Wheatgerm oil.

NZ average diet intake of Vitamin E is 11 gms for men (RDI 10) and 9 mg (RDI 7) for women.

Role in our body
- Assists immune system to kill infections and eliminate poisons.
- Necessary for overall health of blood.
- Anti-oxidant. Protects body cells and fats from oxygen damage.

Too little
- Weak red blood cells (anemia). Low energy.
- Alzheimer's disease (factor).
- Low immunity to infection. Bed sores.
- Increased risk of blood clots resulting in heart attack and stroke.
- Prostate cancer in men.
- Pre-menstrual tension in women (PMT).
- Leg muscle cramps. Growing pains during adolescence.
- Poor blood circulation, especially to legs.

Too much (over 100 mg daily)
- Can increase existing high blood pressure.
- Nausea. Diarrhoea.

Cooking losses
High when nuts are roasted. Heating and freezing is also reported to diminish the Vitamin E content of plant oils but there is limited data available.

HEALTH HINT — **Anti-oxidant supplements**

High priced anti-oxidant supplements have become big business in recent years. Anti-oxidants are essential to our health but there should be no need to pay high prices for them. We can readily obtain all that our body can use from our food, or comparatively cheaper supplements such as Vitamin A, CoQ10, Vitamin C, Zinc and Selenium which are all effective anti-oxidants. The higher concentration often claimed for some bark and seed extracts is largely irrelevant, it is the cost per daily requirement that really matters.

The dramatic cures sometimes quoted by sellers of high priced anti-oxidants were probably individuals whose diets were seriously lacking in anti-oxidants in the first place. There appears to be no evidence of well nourished people gaining any extra benefit from these high priced supplements.

Vitamin E

	RDI	
Men	10.0 mg	
Teenage boys	11.0 mg	
Teenage girls	9.0 mg	
Women	7.0 mg	Breast feeding 9.5 mg

		mg
	BEVERAGES	
5 cups	Tea	1.5
	DAIRY / EGGS	
3 tbsp	Margarine	3.5
1 cup	Soya Milk	2.0
	FRUITS	
½ cup	Blackberries	2.5
½	Avocado	2.0
	GRAINS / NUTS	
1/3 cup	Almonds (raw)	12.5
1/3 cup	Peanuts (raw)	5.0
1/3 cup	Brazil Nuts (raw)	3.5
1/3 cup	Pecan Nuts (raw)	2.5
1 tsp	Peanut Butter	1.5
	MEATS / FISH (cooked)	
100 gms	Eel	5.0
	VEGETABLES (cooked)	
1 med	Kumara	5.0
6 med	Mushrooms	4.0
½ cup	Yams	3.5
1 cup	Taro	3.0
1 bag	Potato Crisps (50gm)	2.7
1 tbsp	Cooking Oil (Soya)	2.5
1 cup	Squash	2.5
½ cup	Asparagus	2.0
2 med	Potatoes	2.0
½ cup	Silverbeet	2.0
½ cup	Spinach	1.5
¾ cup	Baked Beans	1.5
	HEALTH FOODS	
1 tbsp	Wheatgerm Oil	19.0
1 tbsp	Sunflower Seed Oil	7.0
2 tbsp	Sunflower Seeds	6.5
1 tbsp	Cod Liver Oil	3.0
1 plate	Muesli Cereal	3.0
2 tbsp	Wheatgerm	2.5
2 tbsp	Corn/Maize Oil	2.5
1 cup	Tomato Juice	2.2

"If I could live my life over again I would devote it to proving that germs seek their natural habitat, diseased tissues – rather than being the cause of disease."

Dr Rudolf Virchow. 1821-1902 German pathologist.
Father of the 'germ theory' of disease.

CoQ10

This vitamin-like enzyme has anti-oxidant properties similar to Vitamin E. Japanese researchers isolated CoQ10 from plant oils after they discovered that it was almost always deficient in heart attack victims, the aged, and those who suffer from gum disease.

Over 10 million Japanese are now reported to take CoQ10 as a daily supplement, which may explain their superior longevity.

Dr Bliznakov of the Lupus Research Institute of Connecticut reports that 74% of all heart attack patients have been healed when given CoQ10 and that high blood pressure drops markedly.

Elsewhere in an experiment repeated three times, mice treated with CoQ10 lived 56% longer than average.

CoQ enzymes (there are other numbers besides Q10, ie Q7, Q8 and Q9) are relatively abundant in food and a healthy liver will use these other CoQ enzymes to manufacture CoQ10, provided Selenium levels are adequate.

However it does appear as if our liver loses the ability to make CoQ10 as we age for a decline of body levels is usually found as we grow older. When levels drop to 75% of normal, disease appears. When they drop below 25% we die.

Role in our body
- Assists immune system to kill infections and eliminate poisons.
- Anti-oxidant and blood regulator.
- Assists body cells to extract energy from food.

Too little
- High blood pressure. Heart attack. Stroke.
- Gum disease.
- Low energy and weakened muscles.
- Lowered immunity to disease.
- Premature aging and reduced life expectancy.

Too much
No known toxic effects.

Cooking losses
No significant losses.

CoQ enzyme (limited data available)
(Our body uses CoQ to manufacture CoQ10)

RDI 15mg (est)

		mg
	DAIRY / EGGS	
2 med	Eggs Cooked)	6.5
	GRAINS / NUTS (cooked)	
1/3 cup	Peanuts	1.5
	MEAT / FISH (cooked)	
100 gms	Sardines	6.0
100 gms	Pork	3.0
100 gms	Beef	3.0
100 gms	Chicken	2.0
100 gms	Fish	2.0
	VEGETABLES (cooked)	
2 cobs	Sweet Corn	10.0
¾ cup	Baked Beans	6.0
	OILS	
3 tbsp	Soybean oil	3.0
3 tbsp	Rapeseed oil	2.5
	HEALTH FOODS	
1 tbsp	Cod Liver Oil	No figures available but reported to be a rich source.

HEALTH HINT — **Back pain and stress**

A major reason for the sudden onset of back pain, especially when no heavy lifting has been done, is stress.

When we bend our back, many small muscles need to work in sequential order, one after the other. When we are under stress and flustered, the signals from our overloaded brain to these muscles can get out of sequence, resulting in two or more muscles working against one another and tearing slightly. This can result in considerable pain until naturally healed.

Clumsiness and accidents are common when we are in situations that fluster us, such as having to give a speech or back a trailer into a tight spot while others watch. Again because of uncoordinated signals from an overloaded brain.

One way to overcome this stress is to take the time to consciously steady and slow down our breathing.

Vitamin H – Biotin

Biotin is a vitamin manufactured by our bodies in the intestines. Normally deficiencies are not found except after severe dieting, or long term use of antibiotics which kill the bacteria necessary to form the vitamin. Babies are particularly vulnerable.

Raw eggs can prevent biotin from being absorbed from the intestines, but consumption would need to be in the unlikely vicinity of twelve eggs a day to seriously affect an adult. Cooking eggs neutralises the particular protein that causes the problem.

Biotin can also be obtained direct from our food, mainly eggs (cooked), peanuts, liver, malt extract and brewers yeast.

Role in our body
- Helps maintain health of our skin and hair.
- Assists in the manufacture of body fats and acids.

Too little
- Rash around nose and mouth.
- Dry scaly skin. Greyish pallor.
- Stiff hair that is difficult to comb flat. Hair loss.
- Premature greying of hair.
- Low appetite. Nausea.

Too much
No known toxic effects.

Cooking losses
Manufactured by our bodies.

NO FOOD TABLE

"Hunger is good sauce."   Cervantes

HEALTH HINT — After antibiotics

Antibiotics can be effective drugs for killing harmful bacteria and serious infections, but they also kill vitamin producing bacteria in our intestines. Yeast infections and diarrhoea are common disorders following antibiotic treatment.

Yoghurt is an ideal food to help our body re-establish this necessary bacteria during and following antibiotic treatment.

A supplement known as lactobacillus acidophilus is also available from health food shops for this purpose.

HOMEMADE YOGHURT

INGREDIENTS

1 cup	Milk powder (whole or low-fat).
2 cups	Luke warm water.
1 tsp	Honey (optional).
2 tbsp	Yoghurt (not more than ten days old).

Dissolve the honey in the water then add the milk powder and beat. Gently stir in the yoghurt starter and leave overnight in a warm place 35 to 50°C.

Should keep approximately two weeks in a refrigerator.

HEALTH HINT — Avoiding headaches

We can largely avoid headaches by following these four rules:

1. Sleep regular hours.
2. Arise from bed early.
3. Avoid anxious thoughts.
4. Take a daily 2 km walk in the fresh air.

"There is no known disease caused by a deficiency of synthetic pharmaceuticals in the body."   Dr Victor Penzer.

Vitamin K

An oily yellow vitamin that plays a key role in the clotting of our blood. Like Biotin it is able to be manufactured in our intestines, however antibiotics can kill the bacteria necessary to form Vitamin K. Some new born babies are deficient.

The anti-clotting drugs prescribed for some heart attack patients work by hindering the action of Vitamin K, as also does aspirin.

Vitamin K can be obtained from food sources, mainly dark green vegetables such as Silverbeet. Cauliflower is also a good source.

Role in our body
- Assists in the clotting of our blood.
- Helps maintain health of our liver.
- Assists in muscle energy storage.
- Assists in maintaining Calcium in bones.

Too little
- Slow blood clotting. Nose bleeds. Internal bleeding.
- Easy bruising. Purplish skin.
- Liver disorders.
- Nausea in early pregnancy.
- Osteoporosis.

Too much
No known toxic effects of natural Vitamin K. However injected synthetic Vitamin K (new born babies are routinely injected) can sometimes damage red blood cells and cause sweating and chest constrictions.

Cooking losses
Manufactured by our bodies.

NO FOOD TABLE

"Health is a state of complete physical, mental and social well being, not merely the absence of disease and infirmity."
World Health Organisation

Aluminium

Although Aluminium is found in our bodies it is not believed to be a nutrient. It has been included here as it is widely used as a food additive, and in many NZ municipal water treatment stations to clarify cloudy water.

Some of the more common uses of Aluminium are:
- To soften cheese.
- To promote the free running of salt.
- To increase the whiteness of flour.
- As an ingredient in stomach antacid preparations.
- As an ingredient in under-arm deodorants.
- To clarify cloudy water.

Aluminium was long believed to be harmless, but some researchers now claim otherwise. High Aluminium levels have been found in the brains of those who suffered from early senility and identical symptoms have been reproduced in animals by injecting similar levels into their brains.

Also the Aluminium used in stomach antacid preparations has been proven in some cases to cause rapid osteoporosis and painful, weak muscles in the elderly.

Although there is little hard evidence of widespread harmful effects, researchers do suggest that acid foods such as rhubarb not be cooked in Aluminium cookware as Aluminium can be absorbed by such foods.

Role in our body
No known natural role.

Too little
Research indicates the less the better.

Too much
- Inhibits vitamin and mineral action, especially Calcium.
- Linked with early senility.
- Can cause osteoporosis and painful, weak, numb muscles.
- Hyperactivity and attention disorders in children (ADD).
- Fat build-up in kidneys and liver.

Boron

Boron was once thought to be necessary only for plant life, but researchers have found that it plays a role similar to Vitamin D in animal and human health, especially in the reproductive system, brain and bones.

A study released in 2001 found that men with an intake of 2 mg a day had a 62% lower chance of developing prostate cancer compared to those with half that intake. Three earlier studies of humans temporarily deprived of Boron found symptoms similar to that of Lead poisoning, ie poor concentration, weak short term memory, and lack of physical dexterity.

Other overseas studies have shown that the lower the level of Boron in soils, the higher the incidence of osteoarthritis. Jamaican soils are reported to be chronically low, with 70% of the population eventually suffering from arthritis. NZ figures are similar.

Boron is also seriously lacking in many NZ soils, resulting in poor pasture growth and sub-standard farm animal health unless rectified by adding to fertiliser.

The average daily diet intake in NZ is estimated to be about 2 mg, whereas the RDI to provide protection from arthritis is believed to be 3 to 4 mg daily. One Australian doctor reported that 90% of his osteoarthritis patients improved with Boron supplements, most obtaining complete relief from pain.

Boron may also play a critical role in preventing Osteoporosis (the leaching of Calcium from the bones). In one study, twelve older post-menopausal women were supplemented with an additional 3 mg daily of Boron (Sodium Borate). After eight days all of the women markedly reduced excretion of Calcium and Magnesium, and on average doubled their production of oestrogen. Oestrogen is known to protect younger women from Osteoporosis.

Role in our body
- Maintains health of our bones and joints.
- Maintains health of reproductive system.
- Helps maintain health of our brain.

Too little
- Increased risk of prostate cancer.
- Lack of physical co-ordination.
- Weak short term memory. Poor concentration.
- Lack of Oestrogen in women. Osteoporosis.
- Osteoarthritis.

Too much
No known toxic effects below 100 mg daily.

Cooking losses
No significant losses.

Boron

RDI 3 mg (est)		

	FRUITS	mg
1/3 cup	Raisins	2.4
½	Avocado	1.6
1/3 cup	Sultanas	.8
1 cup	Fruit Juice	.6
1	Pear	.5
5	Prunes	.4
2	Plums	.4
5	Dates	.4
5	Apricots dried halves	.4
1	Orange	.3
1	Apple	.3
1	Kiwifruit	.3
	GRAINS	
½ cup	Soybean flour	1.9
½ cup	Oatmeal	.8

	NUTS	mg
1/3 cup	Almond nuts	1.4
1/3 cup	Hazel nuts	1.3
1/3 cup	Brazil nuts	.8
1 tbsp	Peanut butter	.3
	FISH	
100 gms	Fish	.3
	VEGETABLES	
¾ cup	Red Kidney beans	2.8
2 cobs	Sweet Corn	1.4
1 med	Kumara	.3
2 med	Potatoes	.3

Animal nutrition is well provided for by the modern NZ farmer.

Calcium

Lack of this vital bone mineral can have a dramatic effect on our posture and height.

Astronauts have been found to lose around 200 mg of Calcium from their bones each day spent floating in the weightlessness of space. It appears that Calcium migrates from bones that are not being stressed or flexed daily.

This is thought to be one of the major reasons for the widespread incidence of osteoporosis, or gradual loss of bone mass which shrinks the height of the elderly, especially women, making their bones brittle and easily broken. Long confinement in bed and lack of vigorous exercise mimic the effects of weightlessness. Lack of Vitamin D can accelerate the process.

A healthy bone will bend rather than break, like the green branch of a tree as compared to a dead one. Athletic men and women who remain fit and active throughout their lives generally do not develop osteoporosis. Another testimony to life's rule *'Use it or lose it.'*

One reason Osteoporosis is more common in women is put down to the lower activity level of most women, their smaller bone mass, child bearing, and longer life span. However the female hormone oestrogen appears to protect pre-menopausal women to a large degree from this disorder.

Dietary factors also play a role. African women and elderly vegetarians on low protein diets, seldom develop osteoporosis, whereas it is common among Eskimo women on high protein diets. There is a rather disturbing American study which appears to explain this phenomena. A group of normal young men were placed on a controlled diet containing 500 mg of Calcium daily, a low but common intake (the RDI is 800 mg). When they were fed 47 gms of protein daily (RDI 55 gms), 31 mgs (or 6%) of Calcium was absorbed on average by their bodies. When their protein intake was increased to 92 gms daily, no Calcium at all was absorbed. At an intake of 142 gms of protein, 120 mg of Calcium was lost from their bones.

Considering the high intake of protein among New Zealanders (men 105 gms and women 71 gms) this is cause for concern. However it should be explained that our bodies at any time only absorb a portion of the Calcium we consume. This is allowed for in the RDI of 800mg. The net requirement for an average adult is about 125 mg.

Pregnant women and growing children absorb up to 30% and other adults around 25% of total Calcium intake, depending it would seem on how much protein is consumed.

Vegetarians who tend to consume less protein than average were found in one study to absorb and retain a higher proportion of Calcium than non-vegetarians.

High dietary intakes of 1200 mg of Calcium a day are reported to prevent osteoporosis in many cases, but not reverse it.

Lack of Calcium is also a factor in high blood pressure. In one study of 4000 hypertensives, 85% dropped their blood pressure to safe levels by doubling their Calcium intake. Other studies have obtained mixed results, however there is also a complex interrelationship between Potassium and Sodium (Salt). Excess Sodium takes Calcium and Potassium with it as it is expelled from the body. (See section on Potassium.)

One doctor claims 147 health disorders relate to lack of Calcium.

Calcium is widespread in foods and is also found in water. Average intake from tap water is about 11 mg per litre, but can be as high as 200 mg per litre in very hard water.

Our bodies require large amounts of Calcium and milk is still the richest food source, although consumption in NZ has dropped 30% from a high in 1975. However among adults, 64% of Maori, 54% of Pacific Islanders, and 9% of Europeans can have difficulty digesting the casein or lactose in milk, with symptoms of diarrhoea, stomach cramps and excess wind. Children in all these groups have less problems. Yoghurt is more easily digested.

On the other hand, a well fed milk drinker who moves to a country where milk is not generally available, can incur a severe Calcium deficiency until the body adapts to other sources.

The 1997 NZ nutrition survey found widespread suspected deficiencies of Calcium, (only food intake was measured not absorption) – 37% of teenage girls, 33% of teenage boys, 25% of adult women, and 13% of men.

A severe deficiency of Calcium in children can cause identical symptoms to lack of Vitamin D, ie malformed bones, especially the legs. Vitamin D is necessary for the absorption of Calcium.

In one study 80% of women who suffered from PMT had their symptoms vanish or reduce greatly by doubling their Calcium intake.

One in five schizophrenics become normal after Calcium injections. Calcium is a natural tranquilliser which may also partly explain its role in lowering high blood pressure.

Dolomite powder is a concentrated raw form of Calcium (and also Magnesium). When sold as a health food it should have Vitamin D added to aid absorption.

Role in our body
- Forms and maintains health of our bones and teeth.
- Necessary for correct muscle and nerve action and relaxation.
- Helps maintain correct acidity of our blood.
- Assists in the clotting of our blood.
- Helps maintain health of our skin and protection from sunburn.

Too little
- Malformed bones in children.
- Shrinkage of skeleton, brittle easily broken bones (osteoporosis).
- Osteoarthritis. Kidney stones.
- Eye cataracts.
- High blood pressure. Irregular heart beat.
- PMT or PMS in women.
- Irritability. Insomnia. Nervous twitches. Mental disorders.
- Muscle cramps. Numbness. Tingling or trembling (especially legs).
- Increased risk of sunburn and skin cancer.
- Slow blood clotting.

Too much
No known toxic effects of diet excess.

Cooking losses
No significant losses.

"Custom may lead a man into many errors."
Henry Fielding

Calcium

	RDI			
Men	800 mg			
Teenage boys	1200 mg			
Teenage girls	1000 mg			
Women	1000 mg	Pregnancy 1300 mg	Breast feeding 1400 mg	

		mg			
	DAIRY / EGGS			**FRUIT**	
2 cups	Milk (std)	600	½ cup	Rhubarb (stewed)	110
2 cups	Milk (low-fat)	750	1 cup	Pawpaw	100
25 gms	Cheese	170			
½ cup	Ice Cream	100		**FISH**	
	NUTS		100 gms	Sardines (canned)	550
1/3 cup	Almonds	130		**HEALTH FOODS**	
1/3 cup	Brazil Nuts	90	1 tsp	Dolomite Powder	2500
	VEGETABLES (cooked)		1 cup	Soy Milk	250
1 cup	Broccoli	125			
¾ cup	Baked Beans	90			

HEALTH HINT   **Minimise bed rest**

Long term bed rest can do more harm than good. If we are confined to bed for two or more days we should do our best to still exercise as many of our muscles as possible in order to flex our bones and minimise Calcium loss.

We should also get back on our feet as soon as possible to prevent 'rubber knees,' constipation, and the overall weakness that long bed rest causes.

"The doctor of the future will give no medicine, but will interest his patients in the care of the human frame, in diet, and in the cause and prevention of disease."

Thomas A. Edison.

Chlorine

Chlorine in gas form is a deadly poison, fatal to human life at just 1% concentration in air, yet as a liquid it is part of the hydrochloric acid in our stomach that breaks down our food for digestion.

Chlorine is typical of a number of nutrients, harmful in some forms and dosages, but essential for life in others.

Chlorine is added to most domestic tap water supplies and swimming pools to kill bacteria that may be present. It eventually turns into gas and evaporates when exposed to air.

Although sometimes criticised by uninformed people, the chlorination of domestic water supplies has helped to almost eliminate the often fatal water-borne diseases of typhoid and cholera that were common during the nineteenth century.

The unpleasant smell of heavily chlorinated water is sometimes mistaken for Fluoride, but Fluoride is odourless and tasteless in normal concentrations.

Our dietary needs for Chlorine are normally obtained from Salt which is Sodium Chloride (the Chloride portion is 60%), and provided our Salt intake is sufficient, our Chlorine intake should also be adequate.

No RDI has been set for Chlorine as deficiencies are rare and are only found along with severe Sodium deficiency.

Role in our body
- Assists in the digestion of our food.
- Part of our cell fluid.
- Helps maintain health of our lymph system.
- Helps maintain correct acidity of our blood.
- Helps maintain correct action of liver.

Too little
Rare. No known symptoms.

Too much
- Can destroy useful intestinal bacteria and Vitamin E if taken in too concentrated form.

- In gas concentrations of 1% or more can cause fatal suffocation by filling lungs with fluid.

Cooking losses

Boiling food in water can leach out the salt which contains Chloride. This can be recovered if desired, by consuming the water.

HEALTH HINT   **Rid drinking water of Chlorine**

If your local drinking water is heavily treated and has an objectionable Chlorine smell, put some aside overnight in an open container such as a jug. The Chlorine will soon evaporate and the smell disappear. Stirring can speed up the process.

"Hell no! We haven't got that Chlorine muck in our water!"

97

Chromium

Chromium is required by our bodies to process the sugar we consume. The more sugar in our diet, the more Chromium we need.

Fruit sugars and honey have sufficient Chromium, but refined white sugar lacks this important mineral.

Ironically the raw sources of refined sugar are some of the best suppliers of Chromium but it is lost during refining. However it can be reclaimed from molasses, the 'dregs' of refined sugar.

Whole wheat is another good source of Chromium, but 87% is lost when refined to white flour.

New Zealand foods generally test very low for Chromium.

When Chromium is lacking, our blood sugar regulator the pancreas is seriously affected, which can lead to a hypoglycaemic condition – wildly fluctuating blood sugar levels and consequent mood swings. Lack of Chromium is also believed to play a major role in adult-onset diabetes and in heart disease.

When supplements of 200 mcg of Chromium are taken daily, harmful levels of cholesterol have been found to decrease and useful high-density cholesterol to increase.

The 1997 NZ nutrition survey did not include Chromium but American surveys have found widespread deficiencies, some as high as 50% of the population. Older women in particular seem to be at risk. Pregnancy can halve body levels.

Some respected nutritionists also believe that Chromium levels of hospital patients can be seriously affected by intravenous drips of glucose solution given to nourish them after an operation, if Chromium is not included with the solution.

Body levels of Chromium tend to reduce with age in the western world but not in countries with a more natural diet. High supplements of 1000 mcg daily are normally required to restore a deficiency.

Role in our body

- Assists in processing of sugar and alcohol in our body.
- Assists in the manufacture of good cholesterol and insulin.

- Helps regulate cholesterol levels.

Too little
- Loss of effective blood sugar regulation. Hypoglycaemia.
- Adult onset diabetes (factor).
- High cholesterol levels. High blood pressure. Heart attack. Stroke.

Too much
No known toxic effects – difficult to absorb.

Cooking losses
No significant losses.

Chromium (limited data available)

RDI 125 mcg

		mcg			mcg
	FRUITS			**VEGETABLES (cooked)**	
1	Kiwifruit	35	2 med	Potatoes	20
1	Most other fruits (NZ)	5	100 gms	Most vegetables (NZ)	5
	GRAINS / NUTS			**HEALTH FOODS**	
4 slices	Bread - Wholemeal	70	1 tbsp	Brewers Yeast	100
½ cup	Flour - Wholemeal	65			
	MEATS / FISH (cooked)				
100 gms	Scallops	185			
100 gms	Paua	60			

HEALTH HINT — **Use unrefined sugars**

If we need to satisfy a sweet tooth it is better to do so using natural sugars such as fruit or honey, or less refined sugars such as raw sugar, brown sugar, golden syrup, treacle or molasses, all of which contain protective Chromium, rather than foods made with nutrient barren white sugar.

Molasses is especially rich in trace minerals, being at the bottom of the sugar refining process.

Copper

Copper is essential for our brain health and helps preserve the elasticity of our skin, also our arteries, protecting against the effects of high blood pressure.

Yet excess Copper can be as toxic as Lead, Mercury and Cadmium. The tragic death of an Australian baby was traced to Copper poisoning. Mildly acid water (ph4) had been leaching Copper from the water pipes on a farm where the infant's family lived. The baby had been bottle-fed a milk formula using the farm water. The water tested ten times higher than the recommended limit.

A sure sign of high Copper levels in tap water is a blue stain on white porcelain. New Zealand domestic supplies are monitored to prevent excess Copper levels.

Copper is a brain stimulant and a woman's Copper level will normally double during pregnancy to ensure that her baby receives sufficient Copper in the early years when the infant's diet is low in this mineral. The excess Copper in the mother can take three months to return to normal and this is believed to be a factor in the mental instability of some women after pregnancy, although an erratic sleep pattern (2am feeding) is generally accepted as the main cause.

All young mammals, both animal and human normally have higher Copper levels than adults. This is believed by some to explain the liveliness and high energy behaviour of most youngsters. High Copper levels are believed to play a role in the hyperactivity and short attention span of some children or ADD.

The birth control pill can also more than double Copper levels in some women and has been known to bring on schizophrenic-like behaviour.

Vitamin C, Manganese, and Zinc play key roles in regulating our Copper levels. Where an excess of Copper is found, there is usually a deficiency of Vitamin C and Zinc.

A deficiency of Copper is not common, but has been found in intravenously fed hospital patients and where excessive Zinc supplements (over 160mg daily) have been taken.

Role in our body
- Helps to regulate and stimulate our brain.
- Helps protect our nervous system.
- Assists in forming muscle, arteries, veins, skin and bone.
- Helps our body absorb Iron.
- Assists in the pigmentation of hair and skin.

Too little
- Weak arteries and veins, stroke, aneurysm, varicose veins
- Sagging muscles and early wrinkling of skin.
- High blood cholesterol.
- Weak red blood cells (anaemia). Low energy.
- Lack of hair and skin pigmentation.
- Nerve disorders.
- Osteoporosis (factor).

Too much
- Hyperactivity. Insomnia. Anxiety. Mental disorders.
- High blood pressure (anxiety based).

Cooking losses
No significant losses.

HEALTH HINT **Anxiety without obvious cause**

Anxiety without an obvious cause can be the result of a nutritional deficiency or imbalance. High levels of Copper or Lead can bring about such symptoms, but adequate Zinc should flush these excesses from our body.

Lack of any of the B group vitamins, also Vitamin C, Calcium, Magnesium, Manganese and Potassium can produce similar symptoms.

"Happy the man who has a faithful wife, his span of days is doubled."   King Solomon.

Copper

RDI 1500 mcg

		mcg			
	DAIRY / EGGS				
2 cups	Milk (std or low fat)	150	2 cobs	Sweet Corn	380
			½ cup	Green Peas	150
	FRUITS			**FISH (cooked)**	
1 cup	Pawpaw	700	5	Oysters Rock	1800
½	Avocado	275	100 gms	Paua	1040
1	Pear	220	100 gms	Crayfish	900
1 med	Banana	200	5	Oysters Dredged	500
	GRAINS / NUTS			**HEALTH FOODS**	
1/3 cup	Cashew Nuts	830	100 gms	Liver (lamb)	9900
1/3 cup	Brazil Nuts	560	100 gms	Liver (beef)	2300
1/3 cup	Pecan Nuts	460	100 gms	Kidney (beef)	660
1 plate	Cornflakes (with milk)	460	2 tbsp	Pumpkin Seeds	420
½ cup	Rice - Brown (cooked)	340	1 plate	Muesli Cereal	390
1/3 cup	Peanuts (roasted)	270	100 gms	Kidney (lamb)	370
4 slices	Bread - Wholemeal	240	2 tbsp	Sunflower Seeds	310
2	Weetbix (with milk)	200	1 tbsp	Brewers Yeast	300
½ cup	Flour - Wholemeal	190	1 tbsp	Molasses	270
	VEGETABLES (cooked)				
2 med	Potatoes (with skin)	500			
¾ cup	Baked Beans	420			

HEALTH HINT   Copper bracelets and arthritis

78% of arthritics swear they feel less pain when they wear a copper bracelet. Just why they work is not generally known.

"The greatest tragedy that comes to man, is the emotional depression, the dulling of the intellect, and the loss of initiative, that comes from nutritive failure."

Dr James McLester.
Former president of the American Medical Assoc.

Fluoride

A highly controversial mineral that has never been proven necessary to human health, and in excess is harmful to the human body. Fluoride has the effect in small doses however of hardening teeth and bones by increasing their mineral density, but does make them more brittle.

A small proportion of people (one NZ study of children put the figure at 4%) have difficulty in ridding their bones of excess Fluoride. This is commonly the result of weak kidney function.

An optimum intake of 1500 mcg daily was originally claimed in 1962 to reduce tooth decay an average of 57% in children. An eleven year English study ending in 1969, found a 40% reduction in 8 year old children, dropping to 13% at 14 years of age.

However more recent NZ studies, during the 90's, have found decay increasing again among children (probably due to a run down school dental service), and with little difference between fluoridated and non-fluoridated communities. This lack of difference is believed due to the widespread availability of fluoridated toothpaste and higher levels of Fluoride in processed foods.

Fluoride has a delaying action rather than a preventative action on tooth decay. If a child's diet is full of sticky sweet foods, their teeth will still eventually decay, especially molars which trap food in the hollows.

Natural Fluoride is Calcium Fluoride, whereas that added to domestic water supplies is Sodium Fluoride, which is 25 times more easily absorbed by the body. This factor along with the many proven cases of people sensitive to small amounts of Fluoride, plus lack of control over dosage have fuelled the emotional controversy that still continues.

When a water supply is Fluoridated, the dosage is normally calculated on 1 to 1½ litres per adult a day. However labourers toiling in the hot sun have been known to drink up to 12 litres a day. A healthy body will expel the excess through sweat and urination, but as mentioned above, a small percentage of the population can suffer symptoms of poisoning.

Fluoridated toothpaste has been proven effective in providing teeth with the hardening effects of Fluoride. One study showed that Fluoride toothpaste alone can reduce cavities in children around 25%, and save five teeth over a lifetime. But it should not be swallowed. If swallowed, a large glass of milk is an antidote.

Staining of teeth is a reliable sign that a child's intake is too high. This appears as a white and yellowish-brown mottled effect. Mottled teeth are also often brittle and difficult to fill without breaking.

Symptoms of Fluoride poisoning, apart from mottled teeth and bone growths which are permanent, usually clear rapidly in about two weeks when the source of excess Fluoride is removed.

Nutrients play a major role in tooth health, in particular Vitamin A, Vitamin C, Vitamin D, Calcium, Magnesium, Molybdenum and Phosphorus.

Tooth decay is an expensive and major problem in NZ. Severely decayed teeth, besides smelling bad, can also allow entry into the bloodstream of dangerous bacteria and have been linked with Rheumatoid Arthritis.

Role in our body
- Increases mineral density in our teeth and bones.

Too little
- Less resistance to tooth decay and osteoporosis.

Too much (For most people, over 45,000 mcg daily.)
- Mottled, brittle teeth.
- Abnormally hard and brittle bones.
- Headaches.
- Numb or tingly fingers, arms and legs.

Fluoride (limited data available)

RDI Maximum 1000 mcg (not a nutrient).

	BEVERAGES	mcg
4 cups	Tea (untreated water)	1000
4 cups	Tea (treated water)	1900
1 litre	Water (treated)	1100
1 litre	Water (untreated) (avg)	15
	DAIRY / EGGS	
2 cups	Milk (std or non fat)	90
	FRUITS	
1 svg	Most Fruits (avg)	40

	MEATS / FISH (cooked)	
100 gms	Salmon (canned)	330
100 gms	Crayfish	260
100 gms	Fish (avg)	80
100 gms	Meat (avg)	50
	VEGETABLES (cooked)	
1 svg	Most Vegetables (avg)	30

HEALTH HINT — Avoiding tooth decay

If we must indulge ourselves in sticky sweet foods such as chocolate and toffee, which leave a residue in the hollows of our back teeth, we should immediately afterwards chew a dry crust of bread or eat an apple. This should remove the sticky residue from our molars where most decay occurs, but check in a mirror when possible just to make certain it has all gone.

Sugar-acid decay takes place during the first half hour of eating sticky sweets. Brushing our teeth later that night before bed is too late, the damage has been done. Immediate action is required.

HEALTH HINT — Avoiding Asthma

The two main causes of indoor asthma in NZ are airbourne dust-mite faeces from bedding (especially wool blankets) and to a lesser extent wool carpets, and ultra-fine airborne dried saliva from cat fur. Outdoors the main culprit is rye grass pollen.

To prevent indoor asthma, low humidity is essential, mites cannot survive below 50% humidity. The humidity of the average NZ home is typically between 70% and 80%. A dehumidifier or air conditioner can assist in lowering humidity but good ventilation is generally the most effective approach, especially in kitchens when cooking. An over the stove extraction fan is recommended. Bedrooms should be well aired and ideally at least one window open continually.

Also seal fabric mattresses with plastic covers, and wash all bedding, including pillows, every two weeks in hot water over 54°C. Cold or warm water does not kill mites.

Vacuum carpets weekly with the outside house doors wide open to help remove the fine airborne dust and afterward wipe down dust-collecting surfaces with a damp cloth.

Plenty of anti-oxidants from fruit and vegetables and the avoidance of sugary drinks and foods can greatly help prevent oversensitivity to airbourne allergens including rye grass pollen.

Iodine

A tiny but critical amount of this mineral is required by our body to regulate energy and to help control cholesterol levels. It also protects to a degree against harmful effects of radioactive fallout.

Iodine is absorbed from our blood by the thyroid gland located in our neck. If our diet lacks iodine the thyroid gland will enlarge in order to trap more particles. This condition is called goitre and was common in NZ before Iodine was added to table salt. It is beginning to reappear again as table salt usage drops. The 1997 NZ nutrition survey found widespread deficient intakes of Iodine, with average intake, excluding Iodised salt, only 90 mcg a day, a level at which health is at risk. The RDI is 150 mcg for men and 120 mcg for women.

Besides a swollen neck, other symptoms of Iodine deficiency are sluggish obesity, protruding eyes, a thick voice, painful lumps in the breasts of women, and extremely high cholesterol levels.

This last condition can lead to clogged arteries and heart disease.

Most crop growing soils in NZ are deficient in iodine, especially away from the coast. Iodine is abundant in seawater.

Iodine is also a good antiseptic for wounds and can be used by trampers as a water purifier, but this is organic Iodine not the inorganic variety which can be fatal if more than 15,000 mcg are ingested.

Role in our body
- Helps regulate our body energy levels.
- Helps control cholesterol levels.
- Protects against radiation.

Too little (under 50 mcg daily)
- Enlarged thyroid gland (goitre).
- Obesity. Apathy. Thick voice. Severe lack of energy.
- Protruding eyes.
- Malformed and retarded infants if mother is deficient.
- High cholesterol levels. Heart attack. Stroke.
- Sensitivity to cold.

- Senility-like symptoms.
- Polio (factor).
- Dry skin.
- Painful breast lumps in women (Fibrocystic Breast).
- Macular degeneration of the eyes.

Too much (over 1000 mcg daily)
- Temporary acne-like skin rash.
- Temporary suppression of thyroid hormone activity.

Cooking losses
No significant losses.

Iodine (limited data available)

	RDI				
Men	150 mcg				
Women	120 mcg	Pregnancy	150 mcg	Breast feeding	170 mcg

		mcg			
½ tsp	Kelp Powder	1700	2 cups	Milk (std or low fat)	35
100 gms	Fish (avg) (cooked)	200	2 med	Eggs	25
½ tsp	Salt (Iodised table)	104	1 cup	Pumpkin	20
5	Oysters	50	½ tsp	Salt (unrefined)	4

HEALTH HINT — **All required human minerals found in unrefined sea salt**

All trace minerals required by our bodies are found in sea water which contains 77 different minerals and elements. These are in colloidal form and therefore easily absorbed by our body. Malnourished pygmy races are always isolated from ocean food sources.

Some people swear by the health benefits of one tablespoon of sea water taken internally daily. Also sea water sprayed on skin infections, especially acne can promote rapid healing.

Unrefined sea or rock salt (rock salt is sea salt mined from ancient sea beds) is also a good source of sea minerals.

Refined table salt has had all but a few of the 77 minerals and elements removed to provide free running characteristics.

Iron

Researchers have found that there is adequate Iron in most diets but it is absorption of Iron that can be a problem.

Tea and coffee with our meals hinders absorption. One study found that drinking a cup of tea during, or up to 1¼ hours after a hamburger meal, reduced Iron absorption an average of 64%, and one cup of coffee by 39%.

Vitamin C can have the opposite effect. The average absorption of Iron from our food is normally only 10% of that present, which is allowed for in the RDI. However if we consume 25 mg of Vitamin C during a meal, absorption can double to 20%.

Iron deficiencies are usually found among infants, women, growing teenagers, and athletes who lose a lot of sweat. A year 2000 study found 27% of infants to be deficient, and the 1997 nutrition survey found the intake of NZ women of child bearing age to be 42% deficient, older women 26% and 37% of teenagers, who have high needs.

Nevertheless in many instances a low Iron intake can still produce a satisfactory blood test. There is much evidence that a well nourished, healthy person can increase or decrease absorption according to their needs. Pregnant women commonly increase their absorption rate. Pawpaws and pauas are a rich sources.

Iron deficiency leads to blood anaemia, which means that the red cells are weak and cannot carry much energy, resulting in muscle weakness and lack of vitality. This condition can also produce an unnaturally pale skin.

However it is now recognised that other diet deficiencies can mimic this anaemia, especially lack of Vitamin B6 and Zinc which is common and can be more harmful.

Only 1% of NZ men have insufficient Iron intake for normal needs, but too much Iron can be a problem, especially in men over 40. Iron tonics and pills are often found to be the cause.

High Iron levels are found in many arthritics and the birth control pill can raise Iron levels in women.

Role in our body
- Assists in forming of red blood cells.

- Assists in providing energy for our muscles.
- Helps our bodies absorb and utilise protein.
- Helps to build muscle tissue.

Too little
- Weak red blood cells, paleness, low energy (anaemia).
- Muscle weakness, including heart.
- Weakened immune system.
- Fearfulness in children. Decreased attention span. Lack of growth.
- Cracks at the corners of mouth and painful swallowing.
- Yeast infections (Candida).
- Concave nails.
- Poor digestion due to lack of stomach acid.

Too much (over 50 mg daily)
- Greyish hue to skin.
- Arthritis (factor).
- Constipation. Upset stomach.
- Low absorption of Vitamin E.

Cooking losses
No significant losses if cooking fluids retained.

"What makes you think I might be taking too many Iron pills dear?"

Iron

	RDI		
Teenagers	12.0 mg		
Men	7.0 mg		
Women	6.0 mg	Pregnancy	22.0 mg

		mg
DAIRY / EGGS		
2 cups	Milo (all milk)	5.0
2 med	Eggs	2.2
FRUITS		
1 cup	Pawpaw	10.5
1 cup	Cantaloupe	1.4
5	Prunes	1.2
½	Avocado	1.2
GRAINS / NUTS		
1/3 cup	Almond/Cashew Nuts	2.5
½ cup	Flour - Wholemeal	2.2
1 plate	Corn/Wheat cereal	2.2
4 slices	Bread - Wholemeal	2.0
1 tbsp	Cocoa	2.0
100 gms	Chocolate	1.6
1 plate	Porridge	1.5
4 slices	Bread - White	1.2
1 slice	Chocolate Cake	1.2
1/3 cup	Peanuts (roasted)	1.0
MEATS / FISH (cooked)		
1	Paua fritter	11.0
5	Oysters	8.4
100 gms	Mussels	7.7
100 gms	Beef	3.5
100 gms	Lamb	2.7
100 gms	Pork	2.0
100 gms	Chicken	1.9
100 gms	Fish (avg)	1.0
VEGETABLES (cooked)		
¾ cup	Baked Beans	2.9
2 cobs	Sweet Corn	2.4
1 cup	Broccoli	1.6
¾ cup	Peas	1.5
1 bag	Potato Crisps (50gm)	1.4
HEALTH FOODS		
100 gms	Liver/Kidney (lamb)	11.0
100 gms	Liver/Kidney (beef)	8.0
1 cup	Prune Juice	5.0
2 tbsp	Pumpkin Seeds	4.4
1 tbsp	Molasses	3.0
1 plate	Muesli	2.2
1 cup	Soy Milk	2.0
1 tbsp	Treacle	1.8
1 tsp	Marmite	1.8
1 tbsp	Brewers Yeast	1.5
1	Muesli Bar	1.1
2 tbsp	Wheatgerm	1.0

HEALTH HINT — How to absorb more Iron

To ensure good absorption of Iron, teenagers, women and athletes should consume foods containing Vitamin C during most meals, and avoid tea and coffee during, and up to 1¼ hours after a meal.

However if energy levels are high, Iron deficiency is unlikely. Adequate Zinc protects against an excess of Iron.

"The most important crop, is a race of healthy men and women. This is only possible if the soil is fertile."

Sir Albert Howard

Magnesium

In areas of the world where the Magnesium content of drinking water is high (up to 150 mg per litre), there is an average 13% lower incidence of heart disease. Most tap water contains about 2 mg per litre.

New Zealand soils tend to lack Magnesium, an important mineral that helps to hold the Calcium in our bones. It is also believed to protect against Calcium build-up in our arteries. Magnesium deficiencies are common among those with hardened and clogged arteries.

Magnesium is likewise believed to protect against Calcium deposits in our kidneys and urinary system, commonly known as kidney stones.

Some people have found relief from Repetitive Strain Injury (RSI or OOS) when they increased their Magnesium intake.

Our outer tooth enamel is largely Magnesium.

Symptoms of Magnesium deficiency are among the most common health disorders in the western world. This is thought to be because Magnesium is easily lost during the processing of food.

Higher than normal amounts of Magnesium, around 450 mg daily, plus adequate Vitamin B6 have been found to eliminate seizures among many epileptics, and also to relieve painful contractions during late pregnancy.

High dosages as part of Epsom salts (Magnesium Sulphate) are often used as a laxative.

Role in our body
- Helps us absorb and utilise nutrients from food.
- Helps convert blood sugar to energy.
- Assists in correct muscle action.
- Protects and maintains our nervous system.
- Helps maintain correct acidity of our blood.
- Assists in holding Calcium in our bones and teeth.
- Forms outer tooth enamel.
- Helps regulate bone growth in children.
- Helps control cholesterol levels.

Too little
- Depression. Nervousness. Confusion. Epilepsy. Convulsions.
- Lack of co-ordination. Repetitive Strain Injury (RSI or OOS).
- Muscle cramps. Numbness. Tingling. Trembling hands.
- High cholesterol levels. Irregular heart beat. Heart attack. Stroke. Clogged and hardened arteries.
- Calcium build-up (stones) in kidneys and urinary system.
- Prostate swelling.
- Weak bones and easily decayed teeth. Osteoporosis.
- Low energy and tiredness.
- Pre-menstrual tension (PMT).
- Unpleasant body odour.
- Loss of appetite. Nausea.
- Hyperactivity and attention disorders in children (ADD).

Too much
- Diarrhoea.

Cooking losses
No significant losses if cooking fluids retained.

HEALTH HINT   **Stress and heart disease**

Anger and resentment (suppressed or expressed), racing the clock, and prolonged worry can cause clogged arteries and heart disease, just as surely as smoking, poor diet, and lack of exercise.

Three rules to minimise are found in the Health Hint on page 61.

Magnesium

	RDI				
Men	320 mg				
Women	270 mg	Pregnancy	300 mg	Breast feeding	340 mg

		mg			mg
	DAIRY / EGGS			**MEATS / FISH (cooked)**	
2 cups	Milk (std or low-fat)	50	100 gms	Fish (avg)	35
	FRUITS			**VEGETABLES (cooked)**	
1 cup	Pawpaw	170	1 cup	Taro	165
1 med	Banana	55	2 cobs	Sweet Corn	120
	GRAINS / NUTS		¾ cup	Baked Beans	65
1/3 cup	Brazil nuts	210	2 med	Potatoes	35
1/3 cup	Almonds	135	½ cup	Silverbeet	35
1/3 cup	Other nuts (avg)	90		**HEALTH FOODS**	
½ cup	Flour - Wholemeal	60	1 tsp	Dolomite Powder	1500
1 plate	Porridge (with milk)	60	1 tbsp	Molasses	50
4 slices	Bread - Wholemeal	55	2 tbsp	Pumpkin Seeds	160
2	Weetbix (with milk)	55			
½ cup	Rice - Brown	45			

Chronic fatigue syndrome (CFS) linked with low blood pressure

The main cause of Chronic Fatigue Syndrome or CFS (also known as ME or Tapanui Flu) appears to be long term stress combined with an inadequate diet and lack of restful sleep.

Some doctors believe this disorder is nature's way of enforcing rest when we burn the candle at both ends.

It is reported to be notoriously hard to cure quickly, but a combination of nutritious diet, peaceful lifestyle, and patience seems to work in most cases.

Nevertheless, in a 1995 study of 23 sufferers at John Hopkins University it was found that 22 of them had lower than normal blood pressure. They therefore decided to try them on a high salt diet. 19 of the CFS sufferers agreed to this. The result was that nine were completely cured and all of the others improved in varying degrees.

HEALTH HINT — **Breaking an addiction to sleeping pills or tranquillisers**

Until recent years the majority of adults in NZ would at some stage in their lives have been prescribed tranquillisers or sleeping pills to help them through stressful periods. Doctors nowadays are less likely to take this course as many patients rapidly become addicted to these drugs.

For those who are presently addicted and with their doctor's approval would like to give them up, the following guidelines may be helpful.

1. Choose a settled time in which to break the addiction.

2. Be prepared for a temporary increase in anxiety, insomnia and forgetfulness which are common withdrawal symptoms and will soon pass.

3. Cut down your total weekly dosage by approx 25% for the first two weeks, then cut back a further 25% again for another two weeks. Keep reducing the weekly dosage 25% at two weekly intervals until only one pill a week is taken. Then stop completely.

All withdrawal symptoms should end three weeks after the final dose.

"Perfect health is above gold, and a sound body before riches."
King Solomon

Manganese

Manganese is a mineral that has numerous roles in our body and deficiencies seem firmly linked with modern diseases, especially diabetes and artery disease which are often found together. Average NZ intake is generally satisfactory but 10% obtain less than half the RDI.

An American study of 122 diabetics found their average levels of Manganese to be only half those of normal people.

Epilepsy, multiple sclerosis, deafness, balance problems and noises in the ear are other common disorders linked with lack of Manganese.

Epilepsy and multiple sclerosis have been found to respond to Manganese and Zinc supplements, often effecting a cure.

Wheatgerm and bran are rich sources of Manganese but these are mostly removed during the milling of wheat for white flour. Wholemeal flour retains the full amount of Manganese along with other important minerals and vitamins.

Role in our body
- Helps utilise the nutrients from our food.
- Assists in the manufacture and regulation of fats.
- Helps our body manufacture insulin to regulate blood sugar.
- Assists in sex hormone production.
- Assists in forming mother's milk.
- Helps regulate bone growth in children.
- Important for health of brain and nervous system.
- Helps our body eliminate waste and toxins.

Too little
- Nervousness. Insomnia. Mental disorders.
- Convulsions in infants.
- Multiple Sclerosis.
- High cholesterol levels. High blood pressure. Heart attack. Stroke.
- Noises in ears. Balance problems.
- Spinal disc problems.
- Diabetes. Hypoglycaemia.

Too much
Toxic in very high doses causing loss of muscle control and impotency but has only been found in cases of industrial poisoning.

Cooking losses
No significant losses.

Manganese

RDI 3500 mcg

		mcg
	BEVERAGES	
5 cups	Tea	1100
	SWEETS	
100 gms	Chocolate	1300
	GRAINS / NUTS	
1/3 cup	Pecan Nuts	1800
½ cup	Flour - Wholemeal	1800
4 slices	Bread - Wholemeal	1600
2	Weetbix (with milk)	1300
1 plate	Porridge (with milk)	1200
½ cup	Rice - Brown (cooked)	900
1/3 cup	Other Nuts (avg)	800
½ cup	Flour - White	450
4 slices	Bread - White	700

	VEGETABLES (cooked)	
¾ cup	Split Peas	1400
½ cup	Silverbeet	1200
¾ cup	Baked Beans	700
1 med	Kumara	650
2 cobs	Sweet Corn	475
1 cup	Broccoli	450
½ cup	Spinach	400
	HEALTH FOODS	
2 tbsp	Wheatgerm	2300
1 plate	Muesli Cereal	1900
1	Muesli Bar	900
1 tsp	Ginger	500

HEALTH HINT — **Aerobic exercise**

Thirty minutes daily of vigorous, aerobic type exercise (puffing and perspiration), but without excessive jarring of our bones will benefit us more than the questionable claims made for some nutritional supplements.

This kind of exercise may need to be phased in gradually however, especially if our blood pressure is high.

For those who prefer a less strenuous form of exercise, a daily one hour brisk walk can do wonders for our health and mental outlook, also an hourly cat-like stretch is also excellent for our body and can feel totally luxurious when we are in good health.

Molybdenum

Like many of the minerals our bodies require, this lead-like metal also has many industrial uses, from lubrication to the hardening of steel.

Molybdenum is needed by our bodies to help absorb and utilise Copper and Iron. It is also believed to protect our teeth against decay and to assist in regulating uric acid, high levels of which can cause arthritis and gout.

Cancer of the oesophagus (the tube that conveys food to our stomach) is higher than average in areas where the soil is lacking in this mineral.

Molybdenum is found mostly in plant foods, provided the soil is not deficient. New Zealand soils tend to lack Molybdenum.

Lima and Haricot beans are a good food source.

Role in our body
- Helps our body utilise Iron and Copper.
- Helps regulate uric acid levels.
- Assists in sex hormone production.
- Assists in forming tooth enamel.
- Believed to maintain health of oesophagus.

Too little
- Cancer of oesophagus (factor).
- Easily decayed teeth.
- Gout and arthritis.
- Possible Copper and Iron deficiency anaemia.

Too much
- No known toxic effects under 1000 mcg daily.

Cooking losses
No significant losses.

"Nearly all men die of their medicines, not their diseases."

Moliere

Molybdenum (limited data available)

RDI 150 mcg

(Figures vary according to soil content)

		mcg
	BEVERAGES	
1 can	Beer	30
	DAIRY / EGGS	
2 med	Eggs	50
	FRUITS	
½	Rock Melon	30
	GRAINS	
4 slices	Bread - Wholemeal	35
½ cup	Flour - Wholemeal	35
½ cup	Rice - Brown (cooked)	30
	MEATS / FISH (cooked)	
100 gms	Chicken	40
	VEGETABLES (cooked)	
¾ cup	Baked Beans (Haricot)	550
½ cup	Lentils	120
2 cobs	Sweet Corn	120
½ cup	Yams	60
2 med	Potatoes	50
½ cup	Green Beans	40
½ cup	Spinach	25
½ cup	Silverbeet	20
	HEALTH FOODS	
100 gms	Liver (cooked)	200

HEALTH HINT — **Absorbing grains and nuts**

Grains and hard beans require milling, cooking or sprouting before they can be properly digested and the nutrients absorbed. Foods like muesli often contain raw grain that cannot be fully utilised by our bodies.

Similarly nuts require thorough chewing before swallowing. If nuts are swallowed in large chunks they will pass through our digestive system largely undigested.

"The art of living consists of dying young, but as late as possible."
Anonymous

Phosphorus

Phosphorus is the second most required mineral in our body, after Calcium, and is found in every cell. A deficiency of Phosphorus can result in severe tooth decay.

Also critical is the balance of Phosphorus to Calcium,. the normal ratio in our body cells is one part Phosphorus to two parts Calcium. Excess sugar or dietary fat is believed to disturb this balance, leading to weakened bones and teeth. Long term use of stomach antacids can also upset this ratio.

The Phosphorus-Calcium balance is also critical to the absorption of Niacin and in the regulation of the acid-alkaline balance of our blood. If this is not correct arthritis and gout can arise.

Phosphorus tends to be found in high protein foods.

Role in our body
- Helps us absorb nutrients from food.
- Helps maintain the correct acidity of our blood.
- Assists in forming and regulating cells.
- Necessary for correct muscle action.
- Assists in nerve action.
- Assists in the forming of and healthy maintenance of our bones and teeth.

Too little (or wrong ratio to Calcium)
- Malformed bones and teeth and stunted growth in children.
- Weakened bones and teeth. Gum disease. Osteoporosis.
- Arthritis. Gout.
- Cancer (factor).
- Low energy. Muscle weakness. Nervous disorders.

Too much (over 1000 mgs daily)
- Can impair absorption of other minerals.
- Diarrhoea.

Cooking losses
No significant losses.

Phosphorus

RDI 1000 mg

DAIRY / EGGS

		mg
2 cups	Milk (low fat)	600
2 cups	Milk (std)	480
2 med	Eggs	230
25 gms	Cheese	120

GRAINS / NUTS

2	Weetbix (with milk)	270
1 plate	Porridge (with milk)	240
1/3 cup	Nuts (avg)	200
½ cup	Flour - Wholemeal	175
4 slices	Bread - Wholemeal	170
½ cup	Rice - Brown (cooked)	120
4 slices	Bread - White	110

MEAT / FISH (cooked)

100 gms	Sardines (canned)	430
100 gms	Pork	310
100 gms	Beef	240
100 gms	Fish (avg)	230
100 gms	Lamb	220
100 gms	Chicken	215

VEGETABLES (cooked)

2 cobs	Sweet Corn	310
¾ cup	Split Peas	260
6 med	Mushrooms	240
¾ cup	Baked Beans	180
1 cup	Broccoli	130
¾ cup	Peas	105
2 med	Potatoes	105
1 cup	Taro	105

HEALTH FOODS

2 tbsp	Pumpkin Seeds	350
1 plate	Muesli Cereal (with milk)	330
1 tbsp	Brewers Yeast	150
2 tbsp	Wheatgerm	100

HEALTH HINT   The power of affirmations

They may seem foolish at times, but mental affirmations repeated with enthusiasm several times daily have proven to be powerful and effective in swinging our minds from unhealthy negativity to health promoting positiveness very quickly.

Affirmations can be repeated mentally, spoken out loud, or written out.

One all time favourite is *"Every day in every way, I am getting better and better and better."*

Another one to speed recovery to good health is, *"I feel well and healthy, full of energy."*

Potassium

This important mineral works hand in hand with Sodium (Salt) in our bodies, to maintain the correct fluid and electrical balance in our cells and to help control our muscles and nerves.

Recent research has found that it is the depletion of Potassium (and often Calcium) that is the villain in high blood pressure, rather than excess Salt in our diet. Excess Salt by itself does not normally increase our blood pressure, provided our Potassium intake is increased also.

However our reserves of Potassium can be seriously depleted by too much Salt in our diet. For as the excess Salt (which links with Potassium) is expelled from our body, Potassium and Calcium are carried along with it.

Because Salt is so abundant in most NZ diets, Potassium deficiencies are common.

Excess sugar or diuretics also deplete our Potassium reserves. Salt substitutes containing Potassium Chloride can help.

The most common symptoms of Potassium deficiency are sagging weak muscles, excessive loss of body fluids leading to constipation and dry skin, or alternatively, excessive retention of body fluids (edema). Diabetics are often deficient.

We can lose Potassium through excessive perspiration. If we are often in situations where our sweating is heavy, we should make sure that we obtain sufficient Potassium in our diet. Raw fruit is ideal, or fruit juice. But sugary soft drinks, coffee or alcoholic drinks can worsen the situation.

Role in our body
- Helps regulate the fluid balance in our cells.
- Helps maintain the correct acidity of our blood.
- Helps convert our blood sugar to energy.
- Assists our blood to carry oxygen.
- Assists in the formation of muscle tissue.
- Assists in muscle and nerve action.
- Helps regulate growth in children.
- Helps our kidneys eliminate poisons and waste.

Too little (usually due to excess Salt)
- High blood pressure. Stroke.
- Muscle cramps.
- Irregular heartbeat. Weak sagging muscles. Low energy.
- Swelling of legs and other parts of the body (edema).
- Mental apathy and depression.
- Diabetes (factor).
- Constipation.
- Headaches.
- Dry skin.

Too much
No known toxic effects below 18,000 mgs daily if our kidney function is normal.

Cooking losses
Boiling food in water can leach out much of the Potassium but this can be recovered if the cooking water is consumed.

HEALTH HINT   **Five laws of longevity**

A recent extensive study of longevity came to the conclusion that ideally, we should live in a slow-paced society in a mild climate, about 1.5 kms above sea level (1500m), in a region with Selenium-rich soil, and do a lot of walking.

"You can trace almost every sickness, almost every disease to a mineral deficiency."

Dr Linus Pauling (Nobel Prize Winner)

Potassium

RDI 1950 mgs

Qty	Food	mg
	DAIRY / EGGS	
2 cups	Milo (all std milk)	1050
2 cups	Milk - (low fat)	1050
2 cups	Milk - (std)	850
	FRUITS	
1 cup	Honeydew	750
1 cup	Rock Melon	625
1 cup	Cantaloupe	540
1 cup	Pawpaw	495
1 med	Banana	450
½	Avocado	420
1 cup	Orange Juice (pure)	420
5	Prunes	360
5	Apricots (dried halves)	330
¼ cup	Raisins/Sultanas	330
1	Peach	310
5	Dates	310
1	Nectarine	300
1 cup	Grape Juice (pure)	290
½ cup	Grapes	275
1	Kiwifruit	240
1/8	Water Melon	200
	GRAINS / NUTS	
1/3 cup	Nuts (avg)	350
4 slices	Bread - Wholemeal	250
½ cup	Flour - Wholemeal	225
	MEATS / FISH (cooked)	
100 gms	Pork	390
100 gms	Fish	440
100 gms	Beef	330
100 gms	Chicken	320
100 gms	Lamb	275
	VEGETABLES (cooked)	
2 med	Potatoes	980
1 med	Kumara	720
2 cobs	Sweet Corn	720
¾ cup	Baked Beans	600
1 bag	Potato Crisps (50gm)	590
4 slices	Beetroot	440
¾ cup	Taro	400
1 cup	Brussels Sprouts	390
¾ cup	Pumpkin	380
1 cup	Broccoli	360
6 med	Mushrooms	360
1	Tomato (raw)	340
¾ cup	Parsnips	320
1 cup	Cauliflower	260
½ cup	Coleslaw	240
½ cup	Yams	220
¾ cup	Peas	210
1 cup	Cabbage	200
	HEALTH FOODS	
1 tbsp	Molasses	590
1 cup	Soy Milk	320
1 plate	Muesli (with milk)	310
1 tbsp	Treacle	290

HEALTH HINT — **Lowering the risk of stroke**

A 12 year American study of 850 men and women found that just one serving each day of fresh fruit or vegetables (ie high Potassium foods) lowered their risk of stroke 40%.

"It is a marvel to me that we can go on day after day, putting an assortment of objects into a hole in our faces without becoming surprised or bored."   E.M. Forster. British writer.

Selenium

Throughout the world, in every community noted for the longevity of its people, the intake of Selenium is high. Cancer levels are much lower than normal, six times lower in one Finnish study of 12,000 people (increasing to an astonishing eleven times lower when Vitamin E levels were also high). This same study found that heart disease and asthma was much lower and there were fewer birth defects in newborn babies.

Finland like NZ has low levels of Selenium in their crop growing soils. They now add Selenium to crop fertilisers in the form of Sodium Selenate powder and their cancer rate has dropped.

Many Chinese soils also lack Selenium. One Chinese rural community lacking in Selenium took part in what was to be a long term, double blind trial. The residents were given either 200 mcg of Selenium or a placebo. The trial was halted for ethical reasons when cancer rates dropped 33% for those given genuine Selenium.

Selenium along with Zinc and Iodine can be leached out of soils in areas of high rainfall. NZ farmers are aware of this and in many areas, to ensure the health and maximum growth of livestock they add fertilisers containing Selenium to their pastures, or inject it directly into their animals.

The average NZ intake of Selenium is a low 52 mcg daily in the North Island, (it was only 28 mcg a few years ago). The increase is mostly due to imported wheat. However South Island flour is still milled from low-Selenium locally grown wheat.

The Australian RDI for men is 85 mcg which is low by international standards. Most nutrition researchers recommend an RDI of 150 to 200 mcg. One study found that men who received 200 mcg of Selenium daily had a 66% lower risk of prostate cancer.

New Zealanders with very low levels of Selenium are five times more likely to be asthmatics, also flu viruses are often more virulent and can permanently damage heart muscle.

Australian crop growing soils are mostly satisfactory in Selenium and intake across the Tasman is reported to be higher than NZ which may explain why Australia has a 20% lower rate of heart disease than NZ.

In the light of overseas research indicating superior health benefits of an optimum daily intake of Selenium of around 150 to 200 mcg, or 3 mcg per kg of body weight and the current NZ intake of 52 mcg (lower in the South Island), there appears to be sound grounds for insisting that Selenium be added to fertiliser on all deficient crop growing soils in NZ, and pastures used to raise livestock for meat.

For those who wish to have their blood levels of Selenium tested, every 1.0 mcg/ml of Selenium in our blood is roughly equivalent to a dietary intake of 50 mcg for men and 40 mcg for women. At least 2.0 mcg/ml is desirable.

A major role of Selenium in our body is the strengthening of our immune system, especially its anti-oxidant effect which helps protects our cells against the ageing effects of free radical damage. Because of this anti-oxidant effect Selenium is reported to help preserve the elasticity of our skin, and also our arteries, protecting against the effects of high blood pressure.

A further benefit of Selenium unexpectedly reported by participants in an English health magazine test was a notable reduction in anxiety.

In a study of 35,000 American cot deaths, 9000 were found to have a Selenium or Vitamin E deficiency. Vitamin E is also an anti-oxidant and works hand in hand with Selenium. Almost all of the 9000 babies had been bottle fed. Human milk has on average six times as much Selenium, and twice as much Vitamin E as cow's milk.

Selenium is another nutrient that can be toxic in excess (generally over 1200 mcg for several years). In South Dakota, USA grazing animals have been poisoned by natural high levels of Selenium in the soil.

Role in our body

- Anti-oxidant. Assists our immune system to kill cancerous cells, infections and eliminate poisons.
- Helps protect cells and maintain elasticity of tissues and skin.
- Assists in regulating blood pressure.
- Stimulates growth in children.
- Helps maintain health of reproductive system.
- Required for the production of CoQ enzymes.

Too little
- Increased risk of birth defects, cot death and stunted growth in children.
- Low fertility (sperm counts) in men.
- Cystic Fibrosis (clogging of lungs with mucous).
- Low immunity to infection, poisons and all cancers.
- Increased risk of heart attack and stroke.
- Increased risk of asthma.
- Premature ageing of skin and tissues. Skin cancers.
- Brown old age spots on skin.
- Eye cataracts.
- Lower life expectancy.
- Excessive dandruff.

Too much (over 1200 mcg daily)
- Numb or tingly hands and feet.
- Reddish pigmentation to skin.
- Bad breath (garlic-like odour). Metallic taste in mouth.
- Brittle nails, especially thumb nails. Hair loss.

Cooking losses
No significant losses.

"Yes sir, the bacon flavoured tomato project is coming on well. One minor problem ...."

Selenium

	RDI
Men	85 mcg
Women	70 mcg

Figures vary according to soil content.

		mcg
DAIRY / EGGS		
2 med	Eggs	15
2 cups	Milk - (low fat)	6
2 cups	Milk (std or non-fat)	4
FRUITS		
10	Grapes (black)	11
1 cup	Grape Juice (pure)	3
GRAINS / NUTS		
1/3 cup	Brazil nuts	*650
½ cup	Lentils Red	35
4 slices	Bread - Wholemeal	28
½ cup	Flour- Wholemeal	14
1/3 cup	Cashew nuts	12
4 slices	Bread - White	4
1/3 cup	All other nuts	3

* Brazil nuts vary considerably, between 125 mcg and 2650 mcg, according to soil content.

		mcg
MEATS / FISH (cooked)		
100 gms	Snapper	120
100 gms	All other Fish (avg)	50
100 gms	Shellfish (avg)	25
100 gms	Chicken	15
100 gms	Pork	7
100 gms	Beef	4
100 gms	Lamb	4
HEALTH FOODS		
100 gms	Kidney (cooked)	65
¼ cup	Pumpkin seeds	22
¼ cup	Sunflower seeds	21
100 gms	Liver (cooked)	9

HEALTH HINT — Good Selenium sources

It would appear that New Zealanders especially South Islanders may need to make a special effort to obtain sufficient Selenium.

Brazil nuts and snapper are reported to be rich sources, and tablet and liquid supplements are available.

Imported wheat generally contains adequate Selenium, but 85% of this is lost if refined as white flour. Wholemeal flour contains six times the amount of Selenium.

"Medicines are only palliative (symptom-relieving), for back of disease lies the cause, and this cause no drug can reach."

Dr Weir Mitchell.

Sodium (Common Salt)

Sodium is obtained from common salt (Sodium Chloride) and is essential for our good health. Our RDI is about a quarter of a level teaspoon (1000 mg) a day, assuming that we normally drink four cups of liquid. The metric cup measure used throughout this book holds about one and a half tea cups.

Average NZ intake of Sodium is about four times the RDI. However the average Japanese is reported to consume an astonishing thirty times the RDI. Yet their life expectancy surpasses NZ by four years, evidence that we should not approach nutrition too simplistically.

Most dietary salt is concealed in processed foods. Soup mixes and bacon can contain large amounts of salt. Self-raising flour and baking powder are also unexpected sources of Sodium in the form of Sodium Bicarbonate (baking soda).

Most table salt in NZ is refined, to enhance pouring and appearance. This refining is another example of man adulterating healthy food for the sake of convenience. During the refining process of salt over 70 trace minerals and elements are lost, whereas unrefined salt, which can be either evaporated sea salt or rock salt, (rock salt is sea salt mined from an ancient sea bed) contains 77 elements including every trace mineral our body requires for optimum health. Some nutritionists speak of the three deadly white processed foods, white sugar, white flour and white salt. Unrefined salt is grey, chunky and slightly damp and generally needs to be ground before use.

Many health enthusiasts swear by the health benefits of unrefined salt. Some even consume a tablespoon of sea water every day. Healthy long-lived Tibetan people are reported to consume extraordinary amounts of rock salt by stirring it into their tea.

Excess Sodium is eventually expelled from our bodies, but can take valuable Potassium and Calcium along with it, often resulting in a Potassium or Calcium deficiency, both of which are major causes of high blood pressure. Excess salt itself does not normally raise blood pressure if Potassium and Calcium intake is increased also, or surprisingly, if non-Chloride forms of Sodium are used.

Table salt in NZ has Iodine added to compensate for NZ soil deficiencies of this mineral and has proven to be an effective health measure.

Sodium is found in most foods and deficiencies should be rare in normal diets, unless heavy sweating occurs over several days. Lack of Sodium is however becoming more common in NZ as more people switch to a low salt diet for fear of high blood pressure. Consequently Iodine deficiencies are also on the increase.

Symptoms of Sodium deficiency can be leg cramps at night, low energy, aching joints and muscles, dizziness and vomiting. The NZ army issues salt tablets for soldiers on long hot marches.

Role in our body
- Assists in the digestion of our food.
- Assists our body to utilise other minerals, especially Potassium.
- Helps regulate the fluid balance in our cells.
- Helps maintain correct acidity of blood.
- Assists in muscle and nerve action.
- Helps maintain health of our lymph system.
- Helps our body to eliminate waste.

Too little
- Apathy and low energy. Chronic fatigue syndrome (factor).
- Low blood pressure.
- Nerve pains and aching joints and muscles.
- Muscle cramps (especially in our legs).
- Constipation.
- Dizziness and vomiting.

Too much
- High blood pressure (due to loss of Potassium and Calcium).
- Stroke (factor).
- Swelling of tissues (edema).
- Tension. Insomnia.

Cooking losses
Boiling food in water can leach out much of the Sodium. This can be recovered if desired by consuming the water.

Sodium (common Salt)

RDI 1000 mgs (per 4 cups liquid)

BEVERAGES

		mg
1 can	Soft drink (avg)	50
2 tbsp	Milo	45
1 can	Beer	30
2 glasses	Wine	20

DAIRY / EGGS

2 cups	Milk (std or low-fat)	230
2 med	Eggs	160
25 gms	Cheese	150
2 tbsp	Butter	140
2 tbsp	Margarine	140

FRUITS

1 cup	Melons (avg)	24
¼ cup	Dried fruits (avg)	20
1 svg	All other fruits (avg)	4

GRAINS / NUTS

1 plate	Porridge	1480
1 tsp	Baking Powder	900
4 slices	Bread	700
1 svg	Cheesecake	600
½ cup	Flour- self-raising	380
1/3 cup	Nuts - salted	200
1 plate	Wheatflake Cereal (avg)	180
¾ cup	Spaghetti (canned)	170
2 med	Biscuits	60
½ cup	Rice (cooked)	2
1/3 cup	Nuts (unsalted)	2
½ cup	Flour	1

MEATS / FISH (cooked)

100 gms	Bacon	2240
2	Sausages	1400
100 gms	Ham	1400
2	Fish Cakes	1200
1	Hamburger	1200
100 gms	Corned Beef	950
2	Saveloys	900
100 gms	Processed Meat (avg)	900
100 gms	Fish (canned)	600
1	'Oxo' Cube	520
100 gms	Chicken	80
100 gms	Lamb	80
100 gms	Pork (unsalted)	80
100 gms	Fish (avg)	75
100 gms	Beef	50

FAST FOODS

1	Hot Dog	1400
1 svg	Fish and Chips	1300
1	Meat Pie	1050
1	Hamburger (avg)	900
1 svg	Potato Salad	680
1 svg	Pizza	500
1/3 cup	Gravy	400
1 pce	Chicken Fried	350
1 bag	Potato Crisps (med)	325
1 cup	Chips	200
½ cup	Coleslaw	175
1	Fruit Pie	140
1 bag	Potato Crisps (50gm)	85

SALT / SWEETS

1 tsp	Salt - Table	2200
25 gms	Chocolate	30

VEGETABLES (cooked)

1 cup	Soup (avg)	1050
¾ cup	Sweet Corn (canned)	400
½ cup	Peas (canned)	200
4 slices	Beetroot (canned)	200
1 svg	Vegetables (avg)	10

HEALTH FOODS

100 gms	Kidney	300
1 tsp	Marmite/Vegemite	290
1 plate	Muesli	230
100 gms	Liver	150

HEALTH HINT — High sweat loss and salt

If we are sweating continuously in hot conditions and our daily fluid intake has reached four litres, to avoid leg cramps and other problems we should increase our intake of both Sodium and Potassium by 1000 mg for every additional litre of fluid.

Sulphur

Sulphur is one of the lesser talked about minerals, yet there is more Sulphur in our body than salt, and it is found in every cell.

Sulphur is important for the health and appearance of our skin, and also for the gloss and curliness of our hair. A deficiency is known to cause early greying.

Some nutritionists refer to Sulphur as 'nature's beauty mineral.'

Sulphur works hand in hand with protein, and eggs are a good quality source of Sulphur. Other food sources vary in quality.

A respected American nutritionist, Dr Carl Pfeiffer has stated, *"Rheumatoid Arthritis patients seem, as a rule to dislike eggs, at present our only good source of Sulphur. All patients with Rheumatoid Arthritis would do well to eat at least two eggs per day to provide adequate Sulphur for their needs."*

Foods high in Sulphur tend to be smelly when they decompose, fish, meat and eggs are typical.

Role in our body
- Helps our body absorb and utilise nutrients.
- Helps regulate our blood sugar.
- Assists useful bacteria to form in our intestines.
- Helps maintain health of our nervous system.
- Helps maintain health and appearance of our skin and hair.

Too little
- Dry skin.
- Dull or prematurely greying of hair.
- Arthritis.
- Constipation.

Too much
No known toxic effects of dietary Sulphur, however chemical Sulphur can be toxic.

Cooking losses
Up to 50% in vegetables when boiled, but this can be recovered if the water is consumed.

HEALTH HINT — 90 day Arthritis Cure

Currently seven out of ten New Zealanders will eventually suffer from some form of Arthritis.

Dr Joel Wallach NVD, ND is a highly experienced veterinary pathologist turned Naturopath, who has participated in thousands of animal and human autopsies. He lectures widely in the USA on nutritional healing in an outspoken manner and is highly critical of much orthodox drug therapy and surgery.

He claims that all types of Arthritis, even serious bed-ridden cases can be permanently cured in 90 days. Here is the cure:

Take along with normal food, 3 to 6 level tablespoons per day (1 tablespoon per 15kg of body weight) of household Gelatine (which is made from beef cartilage or skin and contains the building material for bone joints), for <u>a full 90 days</u>, plus a <u>complete</u> supplement of high quality dietary minerals in a highly absorbable form, either chelated* or preferably in liquid colloidal* form. Also include 1000mg of Vitamin C daily.

The mineral supplement and Vitamin C can be obtained from health shops.

Gelatine can be ordered in bulk through food supermarkets. For palatability, first dissolve it in boiling water, let it cool and then add fruit juice concentrate to taste.

Gelatine can be replaced with a combination of Glucosamine and Chondroitin (available from health shops) which are said to contain the active ingredients of Gelatine, but reports so far are that this is expensive and not as effective as Gelatine.

Relief may not be noticed until more than a month into the cure. One male reader of the Jan 2001 edition of this book joyfully reported a sudden healing of his painful eight year old spinal arthritis condition around the 85th day of his treatment.

The researcher of this book welcomes feedback from readers on their results from this cure. Feedback to-date (May 2002) has been 100% successful. Not one failure has been reported.

*Chelated means bound with a substance to aid absorption, and colloidal means fine enough to be permanently suspended in water. Sea water and plant juice minerals are colloidal.

"Without health, life is not life, only a state of languor and suffering."
Rabelais, famous French author and physician, 1494-1553.

Sulphur

	RDI
Men	550 mg
Women	450 mg

		mg
	DAIRY / EGGS	
2 med	Eggs	220
2 cups	Milk (low fat)	185
2 cups	Milk (std)	145
	FRUITS	
1 cup	Pawpaw	105
	GRAINS / NUTS	
1/3 cup	Peanuts / Macadamia	190
1/3 cup	Brazil / Cashew	150
4 slices	Bread (all types)	105
1 plate	Porridge (with milk)	90
2	Weetbix (with milk)	90
½ cup	Flour	85
1/3 cup	Almonds	75
	MEAT / FISH (cooked)	
100 gms	Shellfish	470
100 gms	Beef	270
100 gms	Chicken	270
100 gms	Pork	270
100 gms	Lamb	230
100 gms	Fish	220
	VEGETABLES (cooked)	
1 cup	Brussels Sprouts	130
1 cup	Cabbage	90
¾ cup	Baked Beans	90
2 med	Potatoes	80
¾ cup	Split Peas	75
2 cobs	Sweet Corn	75
½ cup	Asparagus	70
1 cup	Cauliflower	60
¾ cup	Peas	55
1 med	Onion	55
	HEALTH FOODS	
1 plate	Muesli Cereal (with milk)	100
2 tbsp	Pumpkin Seeds	80
1 tbsp	Mustard Powder	75

Sulphur, found in eggs, may help to curl our hair . . .

Zinc

This important mineral which is also an anti-oxidant is one of the most seriously deficient minerals in the world today.

As little as ten crops in succession can seriously deplete the readily available Zinc and other minerals from soils, unless replenished by natural river flooding, or added to fertiliser. The soils of Iran, Iraq and Egypt have been seriously deficient of Zinc for centuries. Zinc is not returned to the soil by the Nitrogen-Phosphorous-Potassium (NPK) fertilisers commonly used today. Plants cannot make minerals, they can only extract them from the soil. If the minerals are not in the soil they will not be in the plant. 65% of Zinc is also removed from wheat during refining to white flour.

Nevertheless the 1997 nutrition survey found that NZ intake was surprisingly good, with an average intake for men of 15 mg and women 10 mg. This is due to the high Zinc content of our red meat which tests three times higher than overseas meats. However raw bran with meals has been found to bind Zinc (and other minerals) and hinder it from being absorbed.

The Australian RDI has been lowered from 16 mg to 12 mg in recent years.

Zinc has numerous roles in our body besides being an antioxidant in particular aiding the absorption of other minerals and vitamins and regulating their levels.

Modern health disorders have been strongly linked with low body levels of Zinc. It is believed to be a major factor in heart disease. In Australia low Zinc has been found to be a leading cause of death and stunted growth among Aboriginal children due to their diet changes of recent years.

Other symptoms of Zinc deficiency are repeated infections, slow wound healing, recurrent colds, a poor sense of smell, early greying of hair, thin peeling fingernails with white specks, and stomach skin stretch marks in women after childbirth.

Deficiencies of Zinc are common among bottle-fed babies, women taking the birth control pill, and pregnant women. It is also believed to be a factor in post-natal depression.

Zinc has been used successfully in treating many health

disorders. Senility and eyesight problems have been cured in the elderly, heart disease patients have recovered well, and even dwarfism has responded if caught during the early growth years.

Interestingly, during treatment with Zinc supplements, red-heads have found their hair turning a rich brown, and blondes have become brown haired.

It has been found with some patients that to ensure absorption of a Zinc supplement, it needs to be taken with water on an empty stomach.

In a recent American double blind test involving 580 pregnant Negro women, the half who were supplemented with 25 mg of Zinc during their pregnancy, gave birth to babies an average 4% heavier (mostly bone mass) than the controls. The babies of the lighter women averaged 8% heavier.

NZ Pacific Island women have high levels of Zinc.

Bottle-fed babies have been found to sleep much better when their milk formulae is supplemented daily with 12mg Zinc and 1 mg of Manganese.

Zinc and Vitamin C help our body expel harmful minerals and poisons.

Role in our body
- Anti-oxidant. Helps maintain youthful cells.
- Helps us absorb and utilise nutrients from food.
- Assists in the absorption and regulation of sugar and alcohol.
- Helps maintain the effectiveness of our immune system.
- Helps control our blood cholesterol levels.
- Helps maintain health and correct development of foetus and growing children.
- Helps regulate sexual development in adolescents and maintain fertility in adults.
- Promotes calmness of mind.
- Helps maintain our sense of taste, smell and vision.
- Helps maintain health of our hair and elasticity of our skin.
- Assists in healing wounds and surgery.
- Helps maintain health of the prostate gland in men.
- Assists in the regulation of other minerals and the elimination of toxic metals.

Too little
- Slow wound healing. Infections. Bed sores.
- Yeast infections.
- Poor sense of smell and taste.
- Weak red blood cells, low energy, paleness, (Anaemia}.
- Arthritis, including rheumatoid.
- Prostate gland swelling and cancer.
- Hypoglycaemia. Diabetes.
- Depression, including post-natal. Apathy. Insomnia. Anorexia, Bulimia and other mental disorders.
- Hyperactivity and attention disorders in children (ADD).
- Slow sexual development in adolescents.
- Impotency and low sperm count in males. Infertility in females.
- High blood pressure (anxiety based) and cholesterol levels. Heart attack. Stroke.
- Suppressed immune system. Increased risk of some cancers.
- Stiff or prematurely grey hair and ageing of skin. Falling hair. Excessive dandruff.
- Brittle, peeling nails with white spots.
- Liver disorders.
- Kidney failure (Cadmium poisoning).
- Unpleasant body odours.
- Stretch marks in stomach skin after childbirth.
- Dirt eating (observed among seriously deficient children).
- Poor night vision. Cataracts. Blindness in elderly.
- Dwarfism and mongolism in infants. Lack of growth in children.

Too much (over 100 mg daily)
- Nausea, headache and digestive upsets.
- Lowered blood levels of useful high density cholesterol.
- Copper deficiency.

Cooking losses
No significant losses.

"In health there is liberty. Health is the first of all liberties."

Miel

Zinc

	RDI				
Men	12.0 mg				
Women	12.0 mg	Pregnancy	16 mg	Breast-feeding	18 mg

		mg			
	DAIRY / EGGS				
2 cups	Milk (std or low-fat)	2.0	100 gms	Lamb	4.5
2 med	Eggs	1.6	100 gms	Fish - canned	3.0
			100 gms	Pork	3.0
	FRUITS		100 gms	Crayfish	1.8
1 cup	Pawpaw	1.3	100 gms	Chicken	1.7
			100 gms	Shellfish (avg excl Oysters)	1.6
	GRAINS / NUTS			**VEGETABLES (cooked)**	
1	Bran Muffin	2.8			
1/3 cup	Brazil Nuts	2.1	½ cup	Split Peas	3.3
1/3 cup	Pecan Nuts	2.1	2 cobs	Sweet Corn	2.6
4 slices	Bread - Wholemeal	1.6	¾ cup	Baked Beans	1.5
½ cup	Flour- Wholemeal	1.4		**HEALTH FOODS**	
	MEAT / FISH (cooked)		100 gms	Liver (cooked)	4.0
10	Oysters (rock)	42.0	2 tbsp	Pumpkin Seeds	2.2
10	Oysters (dredged)	16.0	2 tbsp	Sesame Seeds	1.8
100 gms	Beef	6.0	2 tbsp	Wheat Germ	1.6

HEALTH HINT **Colloidal Silver a natural antibiotic**

In recent years the natural antibiotic properties of Silver have been rediscovered. Colloidal Silver (colloidal means fine enough to be permanently suspended in water) either taken internally or sprayed externally has proven effective in killing many stubborn infections without the side effects of pharmaceutical antibiotics.

Colloidal Silver is made by passing a 9 to 30 volt electric current through two pure silver wire electrodes suspended apart in water and can be done at home. The water eventually takes on a slightly gold appearance. There are also machines available in NZ for home use that make the process easy.

"There is always a close connection between violent crime and severe mineral imbalances."

Dr Carl Pfeiffer.
Director Brain Bio Centre, New Jersey.

Conclusions

How marvelous is the human digestive system. How complex and mysterious. One research scientist recently admitted, "We only really understand about 3% of what's going on."

The story is told of Henry Ford's automotive engineers at first giving up in despair at finding a way to cast a V8 engine in one block. When he insisted, they finally devised a way to do it.

Imagine their faces if Henry then said, "OK men, another challenge. We're not going to run this car on petrol, but on vegetables, meat, bread, and water, all sloshed up together in the gas tank. You must design a way for this car to extract its own fuel from those slops. And not only its fuel, but also its engine oil, transmission oil, brake fluid, radiator water, battery acid, rubber to replace what wears off the tyres, and enough carpet and seat leather for a new set every few years."

Impossible for man, yes, but if we think about it, our creator has achieved even more than that with our digestive system.

Good digestion – the key to health

This wonderful digestive system of ours can work reliably for eighty or more years if we just abide by three simple rules:

1. Provide the right raw materials.
2. Provide mental peace while it is doing its work.
3. Wait until hunger tells us it has processed the last lot of food before giving it more (in other words not overeating).

Providing the right raw materials is comparatively easy in NZ. This book is designed to help you do that. The 1997 nutrition survey found that the RDI intake of most New Zealanders was better than expected. In general most vitamins and minerals are supplied in sufficient amounts for good health. Notable exceptions are Folate and Selenium, both important nutrients for our immune systems and many of us will need to go out of our way to obtain the RDI of these.

The proportions of Carbohydrate, Sugars, Protein and Fat are not yet satisfactory but are improving. Eating more natural foods will help.

Rule number two, mental peace while our food is digesting may not be so simple for some of us, but will prevent many illnesses. Anxiety is a major contributor to poor health as it affects our digestion by hindering production of stomach acid and enzymes and consequently the uptake of nutrients. Excess stomach rumbling is a symptom of this common problem. Perhaps we should follow the example of animals and not eat at all when emotionally stressed. Fasting has a powerful calming effect.

The final rule, only eating when genuinely hungry is where most of us sin. Eating more calories than we need is a massive problem in the Western world and one of the major reasons why 75% of us die prematurely from cancer and artery failure, two diseases virtually unheard of in societies that eat natural foods, walk a lot and have a low calorie intake.

Most of us eat three meals a day plus snacks, more because of habit than real calorie need. Perhaps two meals a day without snacks would better suit our needs and help us to keep the most important law of health: "Never eat when not hungry."

Other causes of poor health

There are of course other contributors to ill health. Lack of exercise and obesity can lead to disease, especially western type disease, mainly again because we eat too much and do not expend the calories. Exercise also provides good circulation and perspiration which helps eliminate wastes from our body and it does wonders for our figure, mental outlook and peace of mind.

Tobacco use, excess alcohol and caffeine, illicit drugs, lack of restful sleep, and many prescription drugs slowly but surely undermine our health and interfere with food digestion.

There is little hard evidence that pollutants alone are responsible for much ill health. Our body is highly efficient at expelling poisons, provided we give it good nourishment and a stress-free environment. If we fail to do this then we can suffer severely from common airbourne pollutants that a healthier person will shrug off.

Food allergies

Removing allergenic foods from our diet can sometimes bring about an astonishing recovery from many long standing disorders (even such problems as bed wetting). *'One man's meat is another man's poison.'* People who were bottle fed from birth are much more prone to food allergies.

We can use the Pulse Test to identify a suspected allergenic food. This is done by eating only one type of food each hour, and then measuring our pulse increase (if any), about 30 minutes after taking each food. Any food that consistently increases our pulse by more than 15% is suspect, and should be avoided. It can often be one of our favourite foods.

If you are allergic to wheat you may like to try baking your own bread using untreated flour. Wheat allergies are normally a reaction to the hard-to-digest gluten content of wheat. Extra gluten is added to wheat flours by bakers to increase rising qualities. Your digestive system may be able to handle the lesser amount found naturally in untreated, preferably wholemeal flour. A simple 10 minute bread recipe is found on page 165.

Natural medical treatment

In America and increasingly in Australia, common disorders are being treated very successfully without side effects by doctors who specialise in natural healing using mostly vitamin and mineral supplements. Often in high doses and skilled combinations and sometimes with herbs for heightened effect.

All of the following disorders have been successfully treated:

- Acne
- Alcoholism and drug addiction
- Allergies
- Anaemia
- Arthritis (all types)
- Asthma
- Back Pain
- Bad Breath
- Body Odour
- Boils
- Bronchitis
- Bursitis
- Cancers
- Cataracts
- Clogged Arteries
- Cold Feet
- Constipation
- Cramps
- Diabetes
- Diarrhoea
- Diverticulitis
- Dizziness
- Dry Skin
- Dwarfism
- Eczema
- Edema
- Eye Twitches
- Fibrocystic Breast

- Glaucoma
- Goitre
- Gout
- Gum Disease
- Hair loss (women)
- Hay Fever
- High Blood Pressure
- Hypoglycaemia
- Impotence
- Indigestion
- Infections
- Infertility
- Internal Blood Clots
- Irregular Heart Beat
- Kidney Stones
- Lead Poisoning
- Low Energy
- Macular Degeneration
- Menstrual Difficulties and PMT
- Miscarriages
- Multiple Sclerosis
- Numbness in fingers, toes
- Osteoporosis
- Parkinson's Disease
- Prostate swelling
- Stomach Disorders
- Stunted Growth
- Swollen Legs
- Tooth Decay
- Toxaemia
- Ulcers
- Weak Muscles
- Wind

Also the following mental disorders.

- Convulsions
- Depression
- Epilepsy
- Insomnia
- Irritability
- Panic Attacks
- Paranoia
- Schizophrenia
- Senility
- Weak short-term memory

Vitamin and mineral supplements have also been used successfully with pregnant women where there has been a history of the following disorders in infants of previous births.

- Crooked teeth
- Deformities
- Malformed bones and spine
- Mental retardation
- Mongolism

Dr Ulric Williams – natural healer

Some NZ doctors also prefer to use nutrition as a first approach before resorting to drugs. A pioneer in this field was a New Zealand hospital surgeon and general practitioner named Ulric Williams who became a natural healer.

He originally had the reputation of being a 'playboy doctor' more interested in sports and golf than his patients. He would as one of his golfing friends put it, *"rush them through his surgery, filling them up with sedatives and drugs just as quickly as he could, so that he might have more time for his pleasures."*

One night, at age 42 mid-way through his career, after he had pushed his last patient for the day through and was alone in his surgery, he heard a clear voice speak to him saying, "Are you not ashamed of yourself?"

He believed this voice to be from God and it became a turning point in his life. He rejected what he saw as the hypocrisy of 'the system' and began a career of natural healing. He opened a Health Home in the Wanganui area and began treated chronically ill people who came to him from all over New Zealand. He fed them wholesome food (sometimes just apples and milk for days on end), taught them correct thinking patterns (see page 149) and bullied them into exercising. Results were impressive – almost every one went home healed after a few weeks, even 'terminally ill' cancer patients. Dr Ulric Williams died in 1971 at the age of 81.

Alternative health treatment

Alternative health treatment such as Homeopathy, Herbalism, Osteopathy, Naturopathy, Acupuncture, Bach Flowers, etc, are more gentle than orthodox medicine and have a similar success rate (40%) provided treatment is not left too late. There is also far less danger of medical misadventure and damaging drug side effects. (Interestingly, research shows that doing nothing at all also has an average 40% success rate.)

There are considerable variations in skill however and you may wish to inquire at your local health shop as to the experience and reputation of local natural health practitioners.

Can supplements help us?

The 1997 nutrition survey revealed that about half NZ adults take at least an occasional nutritional supplement, such as Vitamin C tablets. Will supplements like these help us?

If we are having a house built and we supply bricks, nails and tins of paint to tradesmen we have hired, it will not help them unless it is exactly what they need to complete the house. Delivering the same materials to the site without hiring any tradesmen to use them will not help either.

So it is with our bodies, they need both building-block type nutrients, in a form that can be stored until needed, and tradesmen-like working nutrients. Some nutrients serve both purposes and others work as a team, like Vitamin E and Selenium. This book should help you better identify what you are lacking.

However evidence continues to mount that there are no short cuts to health. Just popping a pill rarely works. In one large study reported in this book, synthetic Vitamin E was found ineffective as a supplement, whereas natural Vitamin E obtained from food achieved a massive 62% reduction in heart disease. These are significant life and death figures.

An Australian committee of medical professionals formed to study nutrition and set national RDI's issued this conclusion, *"... fundamental defects in a diet cannot be corrected with vitamin-mineral supplements."*

The speed at which Riboflavin (Vitamin B2) is expelled from our digestive system, as evidenced by bright yellow urine within hours of taking most multi-vitamin pills, strengthens the viewpoint that many nutrients in non-natural form are unable to be utilised effectively by our body.

Riboflavin is the only B vitamin visible in urine, but as all the B vitamins are water soluble and of the same family, it is hard not to draw the conclusion that all the others are being expelled also. Drinking four cups of milk, which contains approximately the same amount of Riboflavin found in an average vitamin pill does not discolour our urine.

Interestingly however, this immediate expelling of Riboflavin and probably the other B vitamins, as confirmed by urine discolouration, does not appear to occur with meal-replacement, milk powder type preparations such as Vitaplan and similar products available from supermarkets, health shops and multilevel marketing distributors, which also contain added Riboflavin and the other B vitamins, nor is it evident with some of the multivitamin pills made in Utah the American centre of natural health products.

Perhaps it is the quality of the ingredients or the pill format that is causing the problem.

In the meantime it would appear safer to first rely on our daily diet to provide the nutrients we need, rather than supplements which may be required but cannot be used in the form supplied.

With the exception of Selenium it should not be too difficult to obtain all that we need from our diet, provided we cut out most of the empty foods and restrict our protein intake to conserve Calcium. An egg-nog for example (egg mixed with milk), contains every nutrient necessary for human life, but not in the ideal proportions (beans and green peas grown in good soil come closest to the perfect natural food).

As mentioned above, Dr Ulric Williams obtained miraculous healings by putting patients on a long term milk and apples diet.

The milk used in his day would not have been homogenised, a modern refining process to lengthen shelf life that many medical researchers condemn as harmful to the body, because of the unnatural fineness of the fat particles.

Natural foods should therefore be our first line of defence. Nevertheless effective vitamin and mineral supplementation might still be required to rectify long standing deficiencies.

Summary

In summary, listed below are six key factors which if observed should enable us to largely maintain excellent health into advanced old age.

Factor one

Our digestive system is a marvellous creation, programmed to grow, heal and maintain our body, all it asks is nutritious food, peace of mind, and eating only when hungry.

Factor two

We must avoid at all costs, long term emotional stresses such as those listed below. These are without doubt the major root cause of adult ill health, principally through poor digestion and a weakened immune system.

Competition	Excitement	Fear	Depression
Worry	Boredom	Grief	Anger
Resentment	Self pity	Despair	

In extreme cases of despair, when the will to live has gone it appears that the human body begins to self-destruct.

On the positive side, a strong will to recover from an illness or injury, coupled with a positive optimistic frame of mind will almost always trigger a strong healing process within our body.

Factor three
Thirty minutes daily of vigorous exercise can improve our health dramatically. Well oxygenated blood heals and energises our body, enlivens our mind and relieves feelings of anxiety, the major cause of poor digestion.

Factor four
Excess refined sugar, and to a lesser extent, excess refined salt can play havoc with our health. Where possible our sugar and salt should be unrefined. When we overcome our childhood-acquired sugar and salt addictions we can safely let our taste buds and appetite regulate our food intake, but we should still keep refined sugars to a minimum to avoid Chromium deficiency. Processed foods are often high in white sugar and refined salt. Ingredients written on packaging are listed in order of weight.

Depression can affect our judgement however.

Factor five
Whole grain products retain most of their nutrients, but refined grains such as white flour and white rice have typically lost 60% to 85% of their vitamins and minerals. <u>The major part of our diet should be natural, unprocessed foods.</u> Generally speaking we should avoid any staple foods that do not 'go off' or attract insects when left on the shelf.

Factor six
Obesity, smoking, the birth control pill, antibiotics and other pharmaceutical drugs, laxatives, antacids, tea and coffee with meals, excess alcohol and caffeine can all play havoc with our health.

Orthodox medical treatment of disease
Orthodox medical advances in the fields of injury treatment and many non-invasive surgical procedures are impressive, and continue to improve.

There are however aspects of the orthodox medical system that appear to be profit driven rather than care driven, and which often result in more harm than good, along with crippling costs to both patients and taxpayers.

Foremost would be the pharmaceutical drug companies, who appear to ignore safe natural remedies in favour of expensive, patentable (ie profitable) chemical drugs and vaccines with inevitable side effects. (See comments by Dr Mendelsohn regarding vaccines, page 148.)

Surgeons also seem to often turn a blind eye to statistics that clearly indicate that the odds of long term improvement with some major surgeries are poor, especially radical heart and cancer surgery (see box page 148).

Although orthodox medicine continues to develop ever more powerful prescription drugs and surgical techniques, they still only attempt to alleviate symptoms or temporarily patch up a problem, rarely do they cure a disorder and restore robust health. Life is often prolonged, but sometimes with much suffering. Often the result is worse than the original disorder. It is obvious to any observant person that drug treatment and major surgeries too often have unpleasant and sometimes fatal side effects.

The American Medical Association reported in 1999 that the third largest cause of death each year in the USA was surgical mis-adventure and fatal side effects of prescription drugs (13% of all deaths). That is five times higher than the American annual road toll which is 2.6% of all deaths (the other major causes of death were alcohol at 4.8%, smoking 7.2%, cancer 22%, and artery disease, ie heart attack and stroke 47%).

NZ statistics are normally comparable to those of America.

Dr Ulric Williams, the surgeon turned natural healer mentioned earlier, was a strong minded, no-nonsense man, constantly in conflict with the orthodox medical authorities of his day. The following blast is taken from one of his many published articles:

"The modern medical system, to the extent of perhaps 80% is nothing but a gigantic, cruel, ludicrous, transparent fraud. Doctors do not know what disease is, nor how it is brought about. They know little of the natural and nothing at all of the spiritual, either for maintaining or regaining good health. They employ a battery of destructive agents more dangerous than the ills they are supposed to cure. Poisonous drugs and vaccines, radiation and mutilating surgery are their weapons."

"Yet principles and methods are available and crying out for recognition so simple and effective that within two months of their adoption perhaps 80% of medics would find themselves out of a

job. That is what they are frightened of, so the entire propaganda machine is operated full blast to make and keep their victims ignorant, sick, terrified and exploitable."

Dr Ulric Williams. NZ surgeon turned natural healer.

Strong words indeed, nevertheless from a doctor who restored robust health to his patients without drugs and surgery.

However we ourselves must accept most of the blame for the emphasis by orthodox medicine on treating symptoms only, and not the cause of disease. We have been reluctant to discipline ourselves and make the simple changes in our lives necessary to maintain good health naturally. We leave things too late, until we are desperate to try anything to lessen our suffering and insist on a quick drug or surgery fix of our unpleasant symptoms.

Dr Ulric Williams once wrote, *"There are no incurable diseases but there are some incurable people, because they don't want to change."*

To use an automotive analogy, treating disease symptoms with drugs that are foreign to our body (except as a last resort) is a little like treating a misfire in a car engine by adding food particles to the petrol. The resultant sluggish performance might disguise the misfire but has not cured it, and performance levels are lower than ever.

To stretch the analogy further, for a surgeon to operate and treat our symptoms only, is like a motor mechanic treating our overheating engine by simply cutting the connection to the water temperature gauge and pronouncing the problem cured.

For genuine healing our body needs the <u>cause of the problem</u> rectified, by changing the lifestyle that caused it in the first place, and by using natural food elements our body is made from, not foreign pharmaceutical chemicals that produce unnatural effects and probably further stress our body in trying to get rid of them.

Common sense also tells us that we should rectify the causes of our health problems early, rather than later.

Effective health care is therefore up to us. We must obey the three rules that govern our digestive system, avoid smoking, exercise daily, and treat ill health early by insisting that medical professionals deal with the causes of our problem, not the symptoms, and as naturally as possible.

Immunising vaccines, proven or unproven?

There are volumes of articles and books both from within and without the medical profession, critical of the dangers and lack of effectiveness of vaccines (a highly profitable commodity for drug companies) and a total absence of convincing responses by the industry.

Vaccines are given credit for healing diseases that were in steep decline long before mass inoculations were introduced. Parents of young children may wish to consider the strong words of Dr Robert Mendelsohn and investigate further.

"I advise mothers to carefully study the known risks of immunisation – Arthritis from German measles shots, encephalitis from measles shots, SID (cot death) following DPT (diphtheria) immunisations, Convulsions from whooping cough vaccine and a host of others."

"There has never been a single vaccine in this country (USA) that has ever been submitted to a controlled scientific study. If you want to be kind, you will call vaccines an unproven remedy. If you want to be accurate you'll call people who give vaccines quacks."

Other concerned doctors have also spoken out strongly.

"Two thirds of the polio cases of this decade (1990's) have been caused by the vaccine itself."
Dr Jonas Salk. Developer of the polio vaccine.

"The result of trials of vaccines show disastrous results, especially in infants." Townsend Letter to Doctors. 1995.

Should cancer always be treated?

"Beyond a shadow of doubt, radical surgery on cancer patients does more harm than good."
Dr Hardin Jones. Dept Medical Physics. University of California.

"Untreated cancer victims tend to live up to four times longer than treated ones." Dr Joseph Beasley, MD.

Healing advice from Dr Ulric Williams

"Hardly anyone would be ill if they hadn't made themselves ill, or been made ill by parents or guardians. The life force within or Spirit of God, is the only power that can make us well. It is always striving to restore us to health and keep us well. It will almost always succeed if we give it a chance and withdraw the physical barriers of over-eating the wrong kinds of food, under-eating the right kinds, insufficient outdoor exercise, and indulging in poisons like tobacco and alcohol. And also the psychological barriers such as fear, resentment, worrying, self pity, jealousy, pride, avarice (greed), gluttony, lust, together with all negative thoughts, beliefs, suggestions, impressions or ideas."

"Of all the causes of disease these psychological barriers are far and away the worst. The diseases will go when we ourselves take the barriers down. They'll be taken away if you let their cause go."

"God is love. Love him and trust in him, and only good and lovely things will happen to us. Health is one of the rewards for living in the right relationship with God."

A personal testimony

The years of research that have gone into this book, now in its seventh edition – the hours of studying massive medical publications, cross referencing, checking, discussing, pondering, attending the occasional lecture, searching the internet, etc, along with four revisions, plus helpful letters and calls from readers over the past twelve years has paid off richly as I have applied the knowledge gained (with some lapses) to improve my own health and that of my family.

Although now in my mid fifties, my energy and fitness level exceeds that of my teens, and I now enjoy total freedom from the regular and miserable bouts of flu and hay fever that once plagued me throughout my life. I rarely get a cold (unless I begin to brag) and touch wood I haven't had one for over two years now.

Apart from Zinc and Selenium I am able to obtain the RDI of all nutrients from regular wholesome food. I take a Zinc supplement every second or third day as I eat little meat, and I add a few drops of liquid Selenium to my homemade wholemeal bread.

My family are not as enthusiastic as myself (they would probably say 'not as fanatical') but their health has also been above average.

Four simple diet rules I try to follow are, 100% wholemeal bread, fruit and vegetables every day, a 2 to 1 ratio of unsaturated to saturated fat, and coming to at least one meal a day hungry. Personally I skip breakfast and find that I never run out of energy before mid-day, even when doing heavy manual work.

A few other general rules, at least half an hour of brisk exercise daily five or six days a week (I like to walk, run, cycle, tramp, row, mow lawns with a hand mower and restore cars), friendly relationships with all, good music, and a driving purpose in life other than retirement.

Yes it is a very true maxim – 'Health is the best Wealth.'

HEALTH HINT — How fit are you?

Research has found that adults who are fit tend to be more alert, positive and productive than their unfit peers. Our resting pulse is usually a good indicator of our fitness.

An approximate guide is as follows:

	Men	Women	
	35	40	Dedicated athlete.
	45	50	Very fit.
Resting pulse	55	60	Fit.
	65	70	Average (ie unfit).
	75	80	Very unfit.
	Over 80	85	See a doctor.

"Some men eat and drink anything put before them, but are very careful about the oil they put in their car." Anonymous

"Those who have no time for exercise will sooner or later have to find time for illness." — Anonymous

Becoming permanently slim

We have an increasing weight problem in NZ – middle aged adult height has remained constant during the eight years between national nutrition surveys 1989 to 1997, but the average weight of middle aged men increased by 3 kg to 85 kg (Maori 87kg, Pacific Islanders 95kg), and women by 4kg to 73kg (Maori 75kg, Pacific Islanders 85kg).

55% of NZ men and 49% of NZ women are overweight.

Dieting alone to lose weight is 98% ineffective – ask any overweight dieter, all that is achieved is a digestive system super-efficient at converting food into fat.

Overweight can be caused by fluid retention due to heart defects, food allergies or vitamin\mineral deficiencies (the fat feels soft and spongy) but this is comparatively rare, excessive calorie intake and lack of exercise are the two main culprits.

The only sure way to lose weight and stay permanently trim (no rolls or flab) is by permanent changes to our lifestyle as outlined in this chapter. The rewards are worth the effort. When we are fit and slim we regain the lively 'full of energy' feeling that we enjoyed as children and our self esteem can soar.

This new lifestyle needs to include the four following essentials:
- Keeping busy.
- Keeping fit.
- Inner peace.
- Controlling calorie intake.

Keeping busy

Boredom is fatal to weight control. If all we have to think about is our next meal, hunger will soon dominate our thoughts. The answer is to fill our waking hours with planned, stimulating work and other interesting activities. Women know how difficult it can be to get a child or husband to come to a meal when they are engrossed in a stimulating task.

TV watching should be minimised – food commercials awaken

our appetite as does also baking cakes and biscuits, etc.

Instead we need to involve ourselves in stimulating activities such as skilled work, education, learning new skills, running a business, investing, travel, sport, discussions, charity work, child care, handcrafts, hobbies, gardening, teaching, etc.

<u>We should wake up every morning to a planned day.</u>

Keeping fit

Fitness is the key to weight control (and food enjoyment). Active children and adolescents are invariably trim, despite huge appetites.

Research has confirmed that our body will normally regulate its optimum weight within narrow limits, once we rise above a certain level of activity and fitness.

The activity necessary to reach this level is:

For vigorous types: 30 minutes daily, five days a week of any energetic activity that will double our normal pulse rate, (normal pulse rate is about 65 per minute for men and 70 for women) ie running, swimming, fast cycling, squash, aerobic groups, gym work, video exercise tapes, home exercise machines, etc.

For less vigorous types: 60 minutes daily of a activity that will increase our pulse rate by 50% or more, ie brisk walking (120 strides a minute), golf, lawn mowing, cycling, tennis, less strenuous video exercise tapes, home exercise machines, etc.

In order to trigger long term weight and fitness effects in our body <u>these exercise periods need to be at least 30 minutes duration.</u> Our body must be convinced that the extra demands we are making on it are likely to be permanent. This is similar to the effect of going barefoot, only if we do it long enough each day will our body get the message and the soles of our feet begin to harden.

Research has found that <u>long term body changes begin to take effect about the 20 minute mark.</u>

Inner peace

Even with the above essentials in place, we can still be defeated if we use food to comfort mental anxieties such as self pity, worry, resentment, etc.

Good nutrition, especially the B Vitamins will help us maintain a serene and cheerful mind, as will vigorous exercise itself. Relaxation techniques can also help.

For serious anxieties, those resulting from deprivation of love, trauma, or sexual abuse during childhood, competent help may be necessary. A breathing therapy commonly known as 'rebirthing,' where childhood trauma is re-experienced under controlled breathing is reported to be highly effective in the hands of an experienced practitioner, and relatively inexpensive. Check the yellow pages under Natural Therapy or ads in Health Magazines. Bach Flower remedies are also reported to be effective.

Peace of mind can often be obtained by simply genuinely forgiving any person whose behaviour has ever hurt us. An attitude of – "I will simply hand it all over to a just and loving God to deal with and get on with my life." Blaming others for our problems will always hold us back in life. A degree of Christian humility is required. Writing a forgiving letter will help greatly, even if it is not sent.

Control calorie intake

Using the calorie charts on pages 19-20 we should work out the average number of calories we consume daily and compare it with the guidelines on page 17.

If we are way above our guideline, merely reducing our calories to the recommended limit may be enough to lose weight long term. This is seldom the case however, overweight people are not usually big eaters, but can be heavy consumers of sweet drinks. Obesity and the constant drinking of sugary drinks seem to go hand in hand. Plain water is far superior.

Excess weight is often the result of binges in past years and we have merely maintained the status quo. The following section on Rapid Weight Loss will show you how to quickly remove the excess, restore your youthful contours, and make a fresh start.

If your calorie intake is close to the limit, provided you exercise sufficiently, your weight should drop to your body's ideal level, ie, no flab or rolls. But if you prefer more rapid progress use the Rapid Loss method.

Changing eating times as outlined in the popular book *"Fit For Life"* has also helped many people, but exercise is a key part of that plan also. We should always remember the first law of health: "Never eat when not hungry."

Breakfast should not be necessary for most adults. Usually we are not hungry at that time of day anyway as it is our body's period for elimination. If we have eaten a full meal the night before and have little genuine appetite in the morning, are in good health and unlikely to be doing a lot of strenuous activity that morning, we can safely wait until lunchtime. (The human body can survive 60 days without food.) We must sit down to at least one meal a day hungry.

Children and adults who do heavy manual work may require a breakfast however. Let appetite be a guide.

Rapid weight loss (3 kg a week)

This rapid weight loss method will remove all your excess weight, no matter how long you have carried it. It works 100% of the time and is probably the most pleasant way possible to seriously lose weight. However it still needs to be followed by the above 'busy, fit and peaceful' lifestyle for maximum benefits.

Only use this method a maximum of two weeks each month to ensure adequate nutrition.

First of all prepare yourself mentally a week before beginning. This is most important. If appropriate let your household know that you will not be eating normal meals next week.

Then using the calorie chart on pages 19-20, or a more detailed one, make up a list of your favourite foods and drinks, in small servings of between 10 and 100 calories each. For example – ½ an egg sandwich - 80 cals, ½ a glass fruit juice - 40 cals, 1 chocolate biscuit - 30 cals, 1 scoop ice cream - 60 cals, ½ cob sweet corn - 70 cals, ¼ of a sausage - 50 cals, and so on.

When the week begins, each day you are to eat three small meals, choosing from the small food servings from your list. The maximum total calories for each day are to be only 400 or less spread over three small meals, ie, 100 to 150 calories a meal. As each meal time comes around, decide what you most feel like eating, within the 100-150 calories allowed.

Include your favourite foods but try and ensure sound nutrition. High fibre vegetable foods provide a full feeling, and carrots which are low in calories are excellent for this purpose. Also whole grains and beans. Minimise white flour and refined sugars.

Now comes the second part, and this is vital, every morning or evening of that week, <u>exercise off all the 400 calories eaten that day.</u> This usually means about 45 minutes of vigorous exercise. Here again do whatever you most feel like doing, running, brisk walking, cycling, or exercising at home doing sit-ups, skips, etc, to lively music or using exercise equipment. The main thing is that you get your heart pumping for at least half an hour and generate enough heat to burn off all (or more) of the 400 calories.

An interesting fact about this 400 calorie method is that you <u>really feel like exercising</u>, far more so than when eating normally.

You must exercise from the first day of the week. The reason for this is that vigorous exercise has the effect of <u>raising your metabolism rate</u>, so that even though you are in effect fasting, you will still stay warm and have plenty of energy. Otherwise your body will go into conservation mode and severely restrict heat generation, which will drastically slow down fat burning.

You should find that exercise becomes effortless as the week goes by, and your weight drops and fitness improves. You will probably also have more mental energy and spiritual sensitivity than normal, and get more done than during a typical week when eating normally.

Dramatic initial weight loss will be mostly fluids, which are just as quickly regained when your calorie intake is increased, but your genuine fat loss should be around 3 kg a week. Also your stomach will shrink and you will be satisfied with smaller meals than normal for the next few weeks.

Carry this method on for a second week if you feel up to it, or eat regular meals for a week and then do it again.

When you have dropped to your optimum weight, no rolls or flab around your waistline and firm upper arms, the four essentials of, <u>Keeping busy, Keeping fit, Inner peace and Controlling calorie intake</u> should never change, lock them into your new lifestyle permanently. Although a few days without exercise won't matter as fitness will usually hold for 10 days before dropping. Nor will the occasional heavy meal do much harm.

Your efforts will be richly rewarded with vigorous good health, a trim body, and perhaps best of all, an alert mind.

Body building

For males who wish to build muscle, this can be achieved by a 30 minute workout, five or six days a week, where the main body muscles are stressed at their maximum limits for at least 15 seconds at a time – ie weight lifting, wrestling, or gymnasium body-building machines.

Do not expect overnight miracles however, normal strength increase and muscle growth averages only 3% a month. However three years continuous work can produce spectacular results on a well proportioned body frame.

HEALTH HINT — **Beat the boredom of exercise**

A popular way to exercise at home is to use a well designed machine such as a walker, exercycle, stepper or rowing machine that allows us to watch TV, listen to tapes or music, read or knit etc, while we are exercising, to take our mind off the effort. Boredom is an enemy to fitness.

It is also a good way to warm up on a cold day.

"This one's been on a fast for three weeks now... "

And God said, "Behold, I have given you every plant, bearing seed... and every...fruit of a tree, yielding seed, to you it shall be for food."   Genesis 1, v 29

Health and religious beliefs

During research for this book, several sources remarked on the superior health of two religious groups – members of the Seventh Day Adventist Church and those of the Church of Jesus Christ of Latter Day Saints (Mormons).

The Seventh Day Adventist lifestyle of vegetarianism and avoidance of alcohol, tobacco and stimulants is based on the writings and teachings of prophetess Ellen White 1827-1915.

The similar Mormon lifestyle, which does permit sparing amounts of meat, is based on a revelation said to have been received from Jesus Christ through the prophet Joseph Smith in 1833, and is known as the 'Word of Wisdom.' It is reproduced on the next page.

A 1989 study of 6000 practising, middle-aged Mormon men by the University of California found an impressive 65% less cancer and 58% less heart disease than non-Mormons, and an average 11 year longer life span.

HEALTH HINT   Fasting and health

One health practice, which may partially explain superior Mormon health, is that of fasting (Mormons fast for 24 hours each month).

Specialist German doctors swear by the health benefits of fasting and they treat disease by supervising fasts of one to three weeks in length, and even up to eight weeks for difficult disorders such as some venereal diseases.

One famous German doctor made this statement, *"There is no disease known to man that cannot be healed by fasting provided the organs are intact."*

"I saw a few die of hunger, of eating, a hundred thousand."
                                                                Benjamin Franklin.

THE WORD OF WISDOM

"Behold, verily, thus saith the Lord unto you – in consequence of evils and designs which do, and will exist in the hearts of conspiring men in the last days, I have warned you, and forewarn you, by giving unto you this word of wisdom, by revelation."

"That inasmuch as any man drink wine or strong drink among you, behold it is not good, neither meet in the sight of your Father, only in assembling yourselves together to offer up your sacraments before him, and behold, this should be wine, yea pure wine of the grape of the vine, of your own make. And again, strong drinks are not for the belly, but for the washing of your bodies."

"And again, tobacco is not for the body, neither for the belly, and is not good for man, but is a herb for bruises, and all sick cattle to be used with judgement and skill."

"And again, hot drinks are not for the body or belly."

"And again, verily I say unto you, all wholesome herbs God has ordained for the constitution, nature, and use of man. Every herb in the season thereof, and every fruit in the season thereof, all these to be used with prudence and thanksgiving."

"Yea, flesh also, of beasts and of the fowls of the air I the Lord have ordained for the use of man with thanksgiving. Nevertheless they are to be used sparingly, and it is pleasing unto me that they should not be used, only in times of winter, or of cold, or famine."

"All grain is ordained for the use of man, and of beasts, to be the staff of life. Not only for man but for the beasts of the field, and the fowls of heaven, and all wild animals that run or creep on the earth, and these has God made for the use of man only in times of famine and excess of hunger."

"All grain is good for the food of man, as also the fruit of the vine, that which yields fruit, whether in the ground, or above the ground. Nevertheless, wheat for man, and corn for the ox, and oats for the horse, and rye for the fowls and for swine, and for all beasts of the field, and barley for all useful animals, and for mild drinks, as also other grain."

"And all saints who remember to keep and do these sayings, walking in obedience to the commandments, shall receive health in their navel and marrow to their bones, and shall find wisdom and great treasures of knowledge, even hidden treasures, and shall run and not be weary, and shall walk and not faint. And I the Lord give unto them a promise, that the destroying angel shall pass by them, as the children of Israel, and not slay them. Amen."

Doctrine and Covenants. (Section 89, verses 4 to 21.)

"When in doubt, try nutrition first."
Dr Rogers, former Dean of Nutrition, age 92 years.

Food additive code numbers

Additives in processed foods have a reputation for being harmful, but 85% of additives have never proved to be detrimental to health in any way.

Additives are normally listed on food packaging, along with the ingredients, in descending order of weight.

Typical additives are anti-oxidants to help prevent fats becoming rancid, baking powder to provide gas for raising dough, colourings to enhance the appearance of food, emulsifiers to allow the mixing of oil and water, starches and gelling agents to thicken liquids, and stabilisers to help prevent ingredients separating.

To identify additives and also to save printing space on food packaging, a code number system has been developed. Below are all the code numbers currently used, and listed alongside each number is the chemical name of the additive, the main use of the additive, and the source. An asterisk * after a code number indicates that adverse reactions have been reported, more often in children.

Foods packaged overseas sometimes have an E before the number, this means that the additive has been approved for use in the European Common Market.

Number, Chemical name, Use, Source
- **100** *Curcumin*, **Yellow colouring**, Turmeric Ginger plant.
- **101** *Riboflavin*, **Yellow colouring**, Yeast.
- **102*** *Tartrazine*, **Yellow colouring**, Synthetic.
- **104*** *Quinoline*, **Yellow colouring**, Coal tar.
- **107*** *Yellow 2G*, **Yellow colouring**, Coal tar.
- **110*** *Sunset yellow FCF*, **Yellow colouring**, Coal tar.
- **120*** *Cochineal*, **Red colouring**, Scale insects.
- **122*** *Carmoisine*, **Red colouring**, Synthetic.
- **123*** *Amaranth*, **Red colouring**, Coal tar.
- **124*** *Ponceau 4R*, **Red colouring**, Coal tar.
- **127*** *Erythrosine*, **Red colouring**, Coal tar.
- **128*** *Red 2G*, **Red colouring**, Coal tar.
- **129** *Allura red AC*, **Red colouring**, Synthetic.
- **132*** *Indigo carmine*, **Blue colouring**, Coal tar.
- **133*** *Brilliant blue FCF*, **Blue colouring**, Coal tar.
- **140** *Chlorophyll*, **Green colouring**, Green plants.
- **141** *Copper phaeophytin*, **Green colouring**, Green plants.
- **142** *Green S*, **Green colouring**, Coal tar.
- **150*** *Caramel*, **Brown colouring and flavouring**, Heated carbohydrates.

151*	*Black BN,*	**Black colouring,** Coal tar.
153*	*Carbon black,*	**Black colouring,** Burnt plants.
154*	*Brown FK,*	**Brown colouring,** Synthetic.
155*	*Chocolate Brown HT,*	**Brown colouring,** Synthetic.
160a	*Carotene,*	**Orange-yellow colouring,** Plants.
160b	*Annatto,*	**Red-yellow colouring,** Annatto seeds.
160c	*Capsanthin,*	**Orange colouring,** Paprika.
160d	*Lycopene,*	**Red colouring,** Tomatoes.
160e	*Apo-8 carotenal,*	**Red-yellow colouring,** Plants.
160f	*Apo-8 carotenoic,*	**Orange-yellow colouring,** Plants.
161	*Xanthopylls,*	**Yellow colouring,** Plants.
161g	*Canthaxanthin,*	**Orange colouring,** Plants.
162	*Betanin,*	**Red colouring,** Beetroot.
163	*Anthocyanins,*	**Red colouring,** Plants.
170	*Calcium carbonate,*	**Firming agent and white colouring,** Natural mineral.
171	*Titanium dioxide,*	**White colouring,** Natural mineral.
172	*Iron oxide,*	**Yellow-red or brown-black colouring,** Natural mineral.
173*	*Aluminium,*	**Silver colouring cake decoration,** Bauxite.
174*	*Silver,*	**Silver colouring cake decoration,** Natural mineral.
175	*Gold,*	**Gold colouring cake decoration,** Natural mineral.
180	*Rubine,*	**Red colouring,** Synthetic.
181	*Tannin,*	**Colour fixer,** Plants.
200	*Sorbic acid,*	**Mould inhibitor,** Mountain Ash berries or synthetic.
201	*Sodium sorbate,*	**Preservative,** Sorbic acid.
202	*Potassium sorbate,*	**Preservative,** Sorbic acid.
203	*Calcium sorbate,*	**Preservative,** Sorbic acid.
210*	*Benzoic acid,*	**Preservative,** Synthetic.
211*	*Sodium benzoate,*	**Preservative,** Benzoic acid.
212*	*Potassium benzoate,*	**Preservative,** Benzoic acid.
213*	*Calcium benzoate,*	**Preservative,** Benzoic acid.
214*	*Ethyl para-hydroxybenzoate,*	**Preservative,** Benzoic acid.
215*	*Sodium ethyl para-hydroxybenzoate,*	**Preservative,** Benzoic acid.
216*	*Propyl para-hydroxybenzoate,*	**Preservative,** Benzoic acid.
217*	*Sodium propyl para-hydroxybenzoate,*	**Preservative,** Benzoic acid.
218*	*Methyl para-hydroxybenzoate,*	**Preservative,** Benzoic acid.
219*	*Sodium methyl hydroxybenzoate,*	**Preservative,** Benzoic acid.
220*	*Sulphur dioxide,*	**Preservative, stabiliser, anti-oxidant and bleach,** Sulphur.
221*	*Sodium sulphite,*	**Preservative and anti-oxidant,** Sulphurous acid.
222*	*Sodium bisulphite,*	**Preservative and bleach,** Sulphurous acid.
223*	*Sodium metabisulphite,*	**Preservative and anti-oxidant,** Sulphurous acid.
224*	*Potassium metabisulphite,*	**Preservative,** Sulphurous acid.
225*	*Potassium sulphite,*	**Preservative,** Sulphurous acid.
226*	*Calcium sulphite,*	**Preservative and firming agent,** Sulphurous acid.
227*	*Calcium hydrogen sulphite,*	**Preservative and firming agent,** Sulphurous acid.
228	*Potassium bisulphite,*	**Preservative,** Sulphurous acid.
234	*Nisin,*	**Preservative,** Bacteria.
235	*Natamycin,*	**Preservative,** Bacteria.
236*	*Formic acid,*	**Preservative,** Synthetic.
249*	*Potassium nitrate,*	**Meat curer and preservative,** Nitrous acid.
250*	*Sodium nitrate,*	**Meat curer and preservative,** Nitrous acid.
251*	*Chile saltpetre,*	**Meat curer and preservative,** Natural mineral.
252*	*Potassium nitrate (saltpetre),*	**Meat curer and preservative,** Natural mineral.
260	*Acetic acid,*	**Preservative and colour diluent,** Synthetic.
261	*Potassium acetate,*	**Preservative and acidity regulator,** Acetic acid.

262	*Sodium diacetate*, **Preservative**, Acetic acid.	
263	*Calcium acetate*, **Preservative and acidity regulator**, Acetic acid.	
264	*Ammonium acetate*, **Preservative**, Acetic acid.	
270	*Lactic acid*, **Preservative, acidity regulator and flavouring**, Milk and plants.	
280	*Propionic acid*, **Preservative**, Fermentation of plants.	
281*	*Sodium propionate*, **Preservative**, Propionic acid.	
282	*Calcium propionate*, **Preservative**, Propionic acid.	
283	*Potassium propionate*, **Preservative**, Propionic acid.	
290	*Carbon dioxide*, **Preservative gas, freezant and propellant**, Natural gas.	
296	*Malic acid*, **Acid flavouring**, Synthetic.	
297	*Fumaric acid*, **Acid flavouring**, Fermentation of plants.	
300	*Ascorbic acid*, **Vitamin C, anti-oxidant and flour improver**, Synthetic.	
301	*Sodium ascorbate*, **Vitamin C and anti-oxidant**, Ascorbic acid.	
302	*Calcium ascorbate*, **Vitamin C and anti-oxidant**, Ascorbic acid.	
303	*Potassium ascorbate*, **Vitamin C and anti-oxidant**, Ascorbic acid.	
304	*Ascorbyl palmitate*, **Vitamin C and anti-oxidant**, Ascorbic acid.	
306	*Tocopherol*, **Vitamin E and anti-oxidant**, Seeds.	
307	*a-Tocopherol*, **Vitamin E and anti-oxidant**, Synthetic.	
308	*y-Tocopherol*, **Vitamin E and anti-oxidant**, Synthetic.	
309	*Tocopherol*, **Vitamin E and anti-oxidant**, Synthetic.	
310*	*Propyl gallate*, **Anti-oxidant**, Tannin.	
311*	*Octal gallate*, **Anti-oxidant**, Tannin.	
312*	*Dodecyl gallate*, **Anti-oxidant**, Tannin.	
315	*Isoascorbate Acid*, **Anti-oxidant**, Ascorbic acid.	
316	*Sodium isoascorbate*, **Anti-oxidant**, Ascorbic acid.	
317	*Potassium isoascorbate*, **Anti-oxidant**, Ascorbic acid.	
318	*Calcium isoascorbate*, **Anti-oxidant**, Ascorbic acid.	
319*	*tert-Butylhydroquinone*, **Anti-oxidant**, Synthetic.	
320*	*Butylated hydroxyanisole*, **Anti-oxidant**, Synthetic.	
321*	*Butylated hydroxytoluene (BHT)*, **Anti-oxidant**, Synthetic.	
322	*Lecithin*, **Stabilizer, anti-oxidant, thickener and emulsifier**, Soya beans.	
325	*Sodium lactate*, **Anti-oxidant ,acidity regulator and moistener**, Lactic acid.	
326	*Potassium lactate*, **Anti-oxidant and acidity regulator**, Lactic acid.	
327	*Calcium lactate*, **Anti-oxidant, acidity regulator and firmer**, Lactic acid.	
328	*Ammonium lactate*, **Stabiliser**, Lactic acid.	
329	*Magnesium lactate*, **Stabiliser and magnesium supplement**, Lactic acid.	
330	*Citric acid*, **Stabiliser, preservative, gelling agent and flavouring**, Molasses.	
331	*Sodium citrate*, **Anti-oxidant, acidity regulator and emulsifier**, Citric acid.	
332	*Potassium citrate*, **Food acid, anti-oxidant and emulsifier**, Citric acid.	
333	*Calcium citrate*, **Emulsifier, firming agent and acidity regulator** , Citric acid.	
334	*Tartaric acid*, **Anti-oxidant, diluent, and food acid**, Grapes.	
335	*Sodium tartrate*, **Anti-oxidant, emulsifier and stabiliser**, Tartaric acid.	
336	*Cream of tartar*, **Anti-oxidant, emulsifier and stabiliser**, Tartaric acid.	
337	*Potassium sodium tartrate*, **Anti-oxidant, emulsifier and stabiliser**, Tartaric acid.	
338	*Phosphoric acid*, **Food acid and anti-oxidant**, Phosphate ore.	
339	*Sodium phosphate*, **Stabiliser and anti-oxidant**, Phosphoric acid.	
339b	*Di-sodium phosphate*, **Stabiliser, nutrient, and gelling agent**, Phosphoric acid.	
340	*Potassium phosphate*, **Stabiliser, anti-oxidant and emulsifier**, Phosphoric acid.	
341	*Calcium phosphate*, **Raising agent, stabiliser, firming agent, anti-caking, anti-oxidant, toothpaste abrasive, nutrient and emulsifier**, Natural mineral.	
341c	*Tri-calcium phosphate*, **Anti-caking, diluent, syrup clarifier, yeast food, and toothpaste abrasive**, Natural mineral.	
343	*Magnesium phosphate*, **Magnesium supplement and anti-caking**, Natural mineral.	
350	*Sodium malate*, **Stabiliser and seasoning**, Malic acid.	

351	*Potassium malate*, **Stabiliser**, Malic acid.	
352	*Calcium malate*, **Stabiliser, seasoning and firming agent**, Malic acid.	
353	*Metatartaric acid*, **Removing calcium from wine**, Tartaric acid.	
355	*Adipic acid*, **Food acid, salt substitute and raising agent**, Synthetic.	
363	*Succinic acid*, **Food acid and gelling agent**, Acetic acid.	
365	*Sodium fumarate*, **Food acid and gelling agent**, Fumaric acid.	
366	*Potassium fumarate*, **Food acid and gelling agent**, Fumaric acid.	
367	*Calcium fumarate*, **Food acid and gelling agent**, Fumaric acid.	
370	*1,4-Heptonolactone*, **Food acid and gelling agent**, Synthetic.	
375	*Niacin (or nicotinamide)*, **Vitamin B3 and colour stabiliser**, Nicotinic acid.	
380	*Tri-ammonium citrate*, **Food acid, emulsifier and stabiliser**, Citric acid.	
381	*Ammonium ferric citrate*, **Iron supplement**, Citric acid.	
384	*Isopropyl citrate mixture*, **Emulsifier and stabiliser**, Citric acid.	
385	*Calcium disodium EDTA*, **Emulsifier and stabiliser**, Synthetic.	
400	*Alginic acid*, **Stabiliser and thickening agent**, Seaweed.	
401	*Sodium alginate*, **Thickening and gelling agent**, Alginic acid.	
402	*Potassium alginate*, **Thickening and gelling agent**, Alginic acid.	
403	*Ammonium alginate*, **Thickening and gelling agent**, Alginic acid.	
404	*Calcium alginate*, **Thickening and gelling agent**, Alginic acid.	
405	*Propylene glycol alginate*, **Emulsifier, thickener and solvent**, Alginic acid.	
406	*Agar*, **Thickening and gelling agent**, Seaweed.	
407	*Irish moss (carrageenan)*, **Thickening and gelling agent**, Seaweed.	
410	*Carob gum*, **Thickening and gelling agent**, Carob beans.	
411	*Oat gum*, **Thickening agent**, Oats.	
412	*Guar gum*, **Thickening agent**, Seeds.	
413	*Tragacanth gum*, **Emulsifier and thickener**, Tree gum.	
414	*Acacia gum*, **Stabiliser, emulsifier, glazing agent and thickener**, Acacia tree gum.	
415	*Xanthan gum*, **Emulsifier and thickener**, Corn sugar.	
416	*Karaya gum*, **Stabiliser, emulsifier and thickener**, Tree gum.	
420	*Sorbitol*, **Sweetening agent and stabiliser**, Glucose.	
421*	*Mannitol*, **Sweetener and anti-caking agent**, Seaweed or plants.	
422	*Glycerol*, **Solvent, moistener, sweetener and lubricant**, Fats and oils.	
430	*Polyoxyethylene*, **Emulsifier and stabiliser**, Synthetic.	
431	*Polyoxyethylene 40*, **Emulsifier**, Synthetic.	
432	*Polysorbate 20*, **Emulsifier and stabiliser**, Sorbitol.	
433	*Polysorbate 80*, **Stabiliser, and moistener**, Sorbitol.	
434	*Polysorbate 40*, **Stabiliser and moistener**, Sorbitol.	
435	*Polysorbate 60*, **Stabiliser, foaming agent, emulsifier and moistener**, Sorbitol.	
436	*Polysorbate 65*, **Stabiliser, defoaming agent, emulsifier and moistener**, Sorbitol.	
440	*Pectin*, **Emulsifier and gelling agent**, Apples and citrus.	
440a	*Ammonium Pectin*, **Stabiliser, gelling agent and thickener**, Pectin.	
440b	*Amidated Pectin*, **Emulsifier, stabiliser, gelling agent and thickener**, Pectin.	
442	*Ammonium phosphatidic acid*, **Stabiliser and emulsifier**, Synthetic.	
444	*Surcrose acetate isobutyrate*, **Stabiliser**, Synthetic.	
445	*Glycerol ester*, **Stabiliser**, Glycerol and wood resin.	
450	*Sodium pyrophosphate*, **Stabiliser, emulsifier and raising agent** , Phosphoric acid.	
450a	*Ammonium phosphate*, **Food acid and yeast activator**, Phosphoric acid.	
451	*Potassium tripolyphosphate*, **Moisturiser and stabiliser**, Phosphoric acid.	
452	*Sodium polyphosphate*, **Stabiliser**, Phosphoric acid.	
460	*Cellulose*, **Fibrous thickener, tablet binder and anti-caking agent**, Plants.	
461	*Methyl cellulose*, **Fibrous thickener, binder and emulsifier**, Wood pulp.	
463	*Hydroxypropyl cellulose*, **Gelling agent**, Cellulose.	
464	*Hydropropyl methyl cellulose*, **Gelling agent**, Cellulose.	
465	*Methyl ethyl cellulose*, **Emulsifier and thickener**, Cellulose.	

466	*Sodium carboxymethyl cellulose,* **Emulsifier, gelling agent and thickener,** Cellulose.	
469	*Sodium caseinate,* **Whitener and emulsifier,** Milk.	
470	*Fatty acids,* **Emulsifier and thickener,** Fats and oils.	
471	*Fatty acids,* **Emulsifier, solvent, stabiliser, lubricant and thickener,** Fats and oils.	
472	*Glycerol acids,* **Emulsifier, solvent and lubricant,** Glycerol.	
472a	*Acetic acid ester,* **Emulsifier, solvent and lubricant,** Acetic acid.	
472b	*Lactic acid ester,* **Emulsifier and stabiliser,** Lactic acid.	
472c	*Citric acid ester,* **Emulsifier and stabiliser,** Citric acid.	
472d	*Tartaric acid ester,* **Emulsifier and stabiliser,** Tartaric acid.	
472e	*Diacetyltartaric acid ester,* **Emulsifier and stabiliser,** Tartaric acid.	
473	*Sucrose esters,* **Emulsifier and stabiliser,** Glycerol and sugar.	
474	*Sucroglyceride,* **Emulsifier and stabiliser,** Glycerol and sugar.	
475	*Polyglycerol acids,* **Emulsifier and stabiliser,** Synthetic.	
476	*Polyglycerol polyricinoleate,* **Emulsifier and fat substitute,** Castor oil and glycerol.	
477	*Propane-1, 2,ester,* **Emulsifier and stabiliser,** Propylene glycol.	
478	*Fatty acid ester,* **Emulsifier and stabiliser,** Lactic acid.	
480	*Dioctyl sodium sulphosuccinate DSS,* **Emulsifier and dissolving agent,** Synthetic.	
481	*Sodium stearoyl lactylate,* **Emulsifier and stabiliser,** Lactic acid.	
482	*Calcium stearoyl lactylate,* **Emulsifier and stabiliser,** Lactic acid.	
483	*Stearyl tartrate,* **Emulsifier and stabiliser,** Tartaric acid.	
485	*Sodium stearyl fumarate,* **Emulsifier and stabiliser,** Fumaric acid.	
486	*Calcium stearyl fumarate,* **Emulsifier and stabiliser,** Fumaric acid.	
491	*Sorbitan monostearate,* **Emulsifier, stabiliser and glazing agent,** Stearic acid.	
492	*Sorbitan tristearate,* **Emulsifier and stabiliser,** Stearic acid.	
493	*Sorbitan monolaurate,* **Emulsifier stabiliser and anti-foaming agent,** Lauric acid.	
494	*Sorbitan mono-oleate,* **Emulsifier and stabiliser,** Oleic acid.	
495	*Sorbitan monopalmitate,* **Emulsifier and stabiliser,** Synthetic.	
500	*Bicarbonate of soda,* **Food alkali, diluent and raising agent,** Synthetic.	
501	*Potassium bicarbonate,* **Food alkali,** Synthetic.	
503	*Ammonium bicarbonate,* **Stabiliser, alkai and raising agent,** Synthetic.	
504	*Magnesium carbonate,* **Alkali, anti-caking and anti-bleaching agent,** Natural mineral.	
507	*Hydrochloric acid,* **Food acid,** Synthetic.	
508	*Potassium chloride,* **Salt substitute, nutrient and gelling agent,** Natural mineral.	
509	*Calcium chloride,* **Stabiliser and firming agent,** Synthetic.	
510	*Ammonium chloride,* **Yeast activator and flavourer,** Synthetic.	
511	*Magnesium chloride,* **Yeast activator,** Synthetic.	
512	*Stannous chloride,* **Yeast activator,** Synthetic.	
513*	*Sulphuric acid,* **Food acid,** Synthetic.	
514	*Sodium sulphate,* **Diluent for colour powders,** Natural mineral.	
515	*Potassium sulphate,* **Low sodium salt substitute,** Natural mineral.	
516	*Calcium sulphate,* **Firming agent, nutrient, stabiliser, yeast activator,** Natural mineral.	
518	*Epsom salts (magnesium sulphate),* **Nutrient and firming agent,** Salt water.	
519	*Cupric sulphate,* **Firming agent and yeast activator,** Natural mineral.	
524	*Sodium hydroxide,* **Colour solvent,** Synthetic.	
525*	*Potassium hydroxide,* **Food alkali,** Synthetic.	
526	*Calcium hydroxide,* **Firming and neutralising agent,** Lime.	
529	*Calcium oxide,* **Food alkali and nutrient,** Limestone.	
530	*Magnesium oxide,* **Food alkali and anti-caking agent,** Natural mineral.	
535	*Sodium ferrocyanide,* **Anti-caking agent,** Synthetic.	
536	*Potassium ferrocyanide,* **Anti-caking agent, removing metals from wine,** Synthetic.	
539	*Sodium thiosulphate,* **Anti-caking agent and chlorine remover,** Natural mineral.	
540	*Calcium hydrogen phosphate,* **Nutrient and raising agent,** Synthetic.	
541	*Sodium aluminium sulphate,* **Raising agent, emulsifier and food acid,** Synthetic.	
542	*Bone phosphate,* **Anti-caking agent, nutrient and tablet base,** Animal bones.	

Code	Name, Function, Source
544	*Calcium polyphosphate*, **Cheese stabiliser**, Synthetic.
545	*Ammonium polyphosphate*, **Water binding aid in meat**, Synthetic.
551	*Silica (silicon dioxide)*, **Suspender, anti-caking agent and thickener**, Sand.
552	*Calcium silicate*, **Anti-caking and non-stick agent**, Natural mineral or synthetic.
553a	*Magnesium silicate*, **Antacid, anti-caking and non-stick agent**, Synthetic.
553b	*Talc (French chalk)*, **Non-stick agent**, Natural mineral.
554	*Aluminium sodium silicate*, **Anti-caking agent**, Natural mineral.
556	*Aluminium calcium silicate*, **Anti-caking agent**, Natural mineral.
558	*Bentonite*, **Filler, emulsifier, clarifier and anti-caking agent**, Clay.
559	*Kaolin*, **Anti-caking agent**, Natural rock.
570	*Stearic acid*, **Anti-caking agent**, Synthetic.
572	*Magnesium stearate*, **Emulsifier, anti-caking and non-stick agent**, Staeric acid.
575	*Glucono delta-lactone*, **Food acid and stabiliser**, Glucose.
576	*Sodium gluconate*, **Neutraliser of adverse minerals, diet supplement**, Gluconic acid.
577	*Potassium gluconate*, **Neutraliser of adverse minerals**, Gluconic acid.
578	*Calcium gluconate*, **Stabiliser, firming agent and neutraliser**, Gluconic acid.
579	*Ferrous gluconate*, **Stabiliser**, Gluconic acid.
620	*Glutamic acid*, **Flavour enhancer and salt substitute**, Fermentation of plants.
621*	*Mono sodium glutamate (MSG)*, **Protein flavour enhancer**, Sugar beets and gluten.
622*	*Mono potassium glutamate*, **Flavour enhancer and salt substitute**, Synthetic.
623*	*Calcium glutamate*, **Flavour enhancer and salt substitute**, Synthetic.
626	*Guanylic acid*, **Flavour enhancer**, Synthetic.
627*	*Disodium guanylate*, **Flavour enhancer**, Synthetic.
629	*Calcium guanylate*, **Flavour enhancer**, Synthetic.
631*	*Disodium inosinate*, **Flavour enhancer**, Protein.
633	*Calcium inosinate*, **Flavour enhancer**, Protein.
634	*Calcium ribonucleotides*, **Flavour enhancer**, Ribonucleic acid.
635*	*Disodium ribonucleotides*, **Flavour enhancer**, Ribonucleic acid.
636	*Maltol*, **Flavour and aroma enhancer**, Plants and synthetic.
637	*Ethyl maltol*, **Sweetener and flavour enhancer**, Maltol.
640	*Glycine*, **Sweetener and flavour enhancer**, Protein.
641	*L-leucine*, **Flavour enhancer**, Synthetic.
900	*Dimethylpolysiloxane*, **Anti-foaming agent and water repellent**, Synthetic.
901*	*Beeswax*, **Glazing and non-stick agent**, Honeycomb.
903*	*Carnuba wax*, **Glazing agent for confectionery**, Wax Palm leaves.
904*	*Shellac*, **Glazing agent for confectionery**, Lac insects.
905	*Mineral oil*, **Glazing, non-stick agent and lubricant**, Synthetic.
920	*L-cysteine hydrochloride*, **Flour improver and flavour enhancer**, Synthetic.
924	*Potassium bromate*, **Oxidiser and flour bleaching agent**, Synthetic.
925*	*Chlorine*, **Antibacterial and flour bleaching agent**, Synthetic.
926	*Chlorine dioxide*, **Antibacterial agent, oxidiser and bleach**, Synthetic.
927	*Azodicarbonamide*, **Flour fermentation improver**, Synthetic.
928	*Benzoylproxide*, **Flour bleach**, Synthetic.
941	*Nitrogen*, **Anti-oxidant and propellant**, Natural gas.
942	*Nitrous oxide*, **Aerator and propellant**, Synthetic.
943	*Butane*, **Propellant**, Petroleum.
944	*Propane*, **Propellant**, Petroleum.
950	*Acesulphame potassium*, **Intense sweetening agent**, Synthetic.
951*	*Aspartame*, **Intense sweetening agent**, Synthetic.
952*	*Cyclamic acid*, **Sweetening agent**, Synthetic.
954	*Saccharin*, **Intense sweetening agent**, Synthetic.
957	*Thaumatin*, **Intense sweetening agent**, Plant.
965	*Glucose syrup*, **Sweetening agent**, Plants.
967	*Xylitol*, **Sweetening agent**, Wood pulp.

1100 *Amylase*, **Enzyme to convert starch to sugars**, Natural enzyme.
1101 *Protease*, **Enzyme to convert protein to sugars**, Natural enzyme.
1103 *Invertase*, **Enzyme to convert sugars to glucose or fructose**, Natural enzyme.
1104 *Lipase*, **Enzyme to convert fats to sugars**, Natural enzyme.
1200 *Polydextrose*, **Bulking agent, calorie reducer and texture regulator**, Synthetic.
1400-1450 *Modified starches*, **Thickening agents**, Plants.
1520 *Propylene glycol*, **Anti-freeze**, Plants.

10 MINUTE WHOLEMEAL BREAD

(Makes two tasty, healthy loaves with a crisp crust.)

INGREDIENTS

6½ cups *	**Freshly ground wholemeal flour.**
3 tsp (raised)	**Baking soda.** (Stir into the water.)
½ cup	**Soya oil or equivalent.**
½ cup	**Treacle or brown sugar.**
1 tsp (level)	**Unrefined salt.** (If using refined salt halve the quantity.)
4 cups	**Water.** (Amount may need to be varied according to type of flour.)

* Use a common household mug as a cup measure.

Stir all ingredients together in a large bowl with a sturdy wooden spatula until you have an elastic dough that will hold its shape without slumping (approx 2 minutes). Kneading is not necessary.

Transfer the dough into two teflon-coated or greased loaf tins and bake for approx 1 hour (depending on darkness or crispness of crust desired) at 200°C (400°F).

For a crisp all over crust, remove the loaves from the tins while still hot.

Index

Acid stomach how to prevent 72
Acne lack of Vitamin B6 59; Niacin can treat 55; Vitamin A used to dry 46; sea water can rapidly heal 107
Acupuncture similar success rate to orthodox medicine 142
ADD See **Hyperactivity, Attention disorder**
Addictions lack of Vitamin C 75
Affirmations powerful and effective 120
Aggressiveness sign of hypoglycaemia 44
Aging premature lack of CoQ10 84; lack of Selenium 126; lack of Vitamin C 76
Alcohol section on 15; excess can raise cholesterol levels 25
Alcoholics 95% incidence of hypoglycaemia among 44; commonly lack Folate 62; often deficient in Thiamine 49
Alcoholism lack of Niacin 56; Successfully treated naturally 140
Allergies can cause fluid retention weight problem 151; dramatic fall-off at close of World War 2 16; excess calories can cause 16; excess sugar can increase sensitivity to 45; lack of Vitamin B12 66; lack of Vitamin C 76; successfully treated naturally 140; those bottle fed from birth more prone to 140; wheat allergy normally reaction to gluten content 140
Alternative health treatment similar success rate to orthodox medicine 142
Aluminium section on 89
Alzheimer's disease high cholesterol linked to 26; lack of Folate 62; lack of Vitamin E (factor) 82; signs of 67
Anaemia lack of copper 101; lack of Folate 63; lack of Vitamin B12 66; lack of Vitamin B6 59; lack of Vitamin C 76; lack of Vitamin E 82; lack of Iron 109; lack of Zinc 136
Aneurysm lack of copper 101
Anger can only exist with blame 35; major cause of ill health 35
Anorexia lack of Niacin 56; lack of Thiamine 50; lack of Vitamin B6 59; lack of Zinc 136; Vitamin A levels high in group of anorexic girls 46
Antibiotics can kill useful bacteria in intestines 86; yoghurt helps body re-establish necessary bacteria following 87
Anti-oxidant help prevent fats becoming rancid 159; CoQ10 similar to Vitamin E 84; Vitamin A an important 46; Vitamin C an important 74; Vitamin E powerful 81; Selenium a major 126; Zinc 134; no need for high priced 82
Anxiety Bach Flowers effective in treating 153; dramatic fall-off at close of World War 2 16; excess calories can cause 16; excess copper 101; lack of Inisitol 69; lack of Niacin 55; lack of Vitamin B6 59; lack of Vitamin C 75; methods of calming 153; sign of hypoglycaemia 44; without obvious cause 101

Apathy lack of Iodine 106; lack of Potassium 122; lack of Sodium 129; lack of Thiamine 50; lack of Zinc 136
Appetite loss of lack of Magnesium 112; lack of Pantothenic Acid 71; lack of Thiamine 50
Appetite low lack of Biotin 86; lack of Vitamin B6 59; lack of Niacin 56
Apples oxidation damage seen in browning 81; fed them apples and milk 142
Arms numb or tingly excess Fluoride 104
Arteries clogged section on Artery Disease **23**; excess saturated fat in diet 34; lack of Folate 63; lack of Magnesium 112; prolonged stress can cause 26; successfully treated naturally 140
Arteries hardened lack of Magnesium 112
Arteries weak lack of Copper 101
Artery disease See **Arteries clogged**
Arthritis 90 day cure 132; Copper bracelets reduce pain 102; emotional stress a key factor 71; German measles shots 148; lack of Boron 91; lack of Calcium 94; lack of Folate 62 63; lack of Molybdenum 117; lack of Sulphur 131; lack of unsaturated fat 34; lack of Vitamin B6 59; lack of Vitamin C 76; lack of Zinc 136; successfully treated naturally 140; wrong Phosphorus-Calcium balance 119
Arthritis rheumatoid eat two eggs a day to provide Sulphur 131; lack of Pantothenic Acid 71; lack of Zinc 136; tooth decay linked with 104
Aspirin destroys Vitamin C in blood 75
Asthma excess sugar can increase sensitivity to 45; how to avoid 105; lack of Selenium 126; much lower when Selenium and Vitamin E levels high 124; NZ'ers with low Selenium five times more likely to suffer from 124; successfully treated naturally 140
Athletes perform better on carbohydrates 21
Attention disorder excess Aluminium 89; lack of Folate 63; lack of Iron 109; lack of Magnesium 112; lack of Vitamin B6 59; lack of Thiamine 50; lack of Zinc 136

Bach Flowers effective in treating anxiety 153; similar success rate to orthodox medicine 142
Back pain stress can cause 85
Bad breath how to banish 80; lack of Vitamin C 76; rotting teeth or infected tonsils 80
Baking powder can destroy Riboflavin 53
Baking soda can destroy Pantothenic Acid 72
Balance problems lack of Vitamin B12 66; lack of Manganese 115
Beans close to perfect natural food 144
Beans baked Haricot, figures for Lentils, Mung, Soya and Lima beans similiar 14
Bed rest can do more harm than good 95
Bed sores lack of Vitamin C 75; lack of Vitamin E 82; lack of Zinc 135
Bed wetting can be caused by food allergy 140

Biotin section on **86**
Birth control pill can lower Vitamin B6 levels 58; can lower B12 levels 66; can more than double Copper levels 100; nutrients to supplement 60
Birth defects excess Vitamin A 47; fewer when Selenium and Vitamin E levels high 124; See also **Cleft palate, Malformities**
Bleeding internal lack of Vitamin C 76; lack of Vitamin K 88
Blood sticky and clots easily when blood sugar high 44; well oxygenated blood heals and energises body 145
Blood acid-alkaline Phosphorus-Calcium balance critical 119
Blood poor circulation lack of Vitamin E 82
Blood clots lack of unsaturated fat 34; lack of Vitamin E 82; risk higher following rich fatty meal 31; successfully treated naturally 141
Blood clotting slow excess unsaturated fat 34; lack of Vitamin K 88
Blood fats Vitamin C can lower 40% 74
Blood pressure guidelines for 28
Blood pressure high section on Artery Disease 23; excess sugar 44; CoQ10 can reduce markedly 84; excess Sodium (factor) 129; fibre can lower by thinning blood 37; how to avoid 35; lack of Calcium 94; lack of Chromium 99; lack of CoQ10 84; lack of fibre 38; lack of Manganese 115; lack of Niacin 56; lack of Potassium 122; lack of Vitamin C 75; prolonged stress can cause 26; successfully treated naturally 141; Vitamin C can lower 74; years lost by (chart) 28
Blood pressure high (anxiety based) excess copper 101; lack of Zinc 136; study of confronters and avoiders 29
Blood pressure low lack of Sodium 129; linked with Chronic Fatigue Syndrome 113
Body odour lack of Magnesium 112; lack of Zinc 136; successfully treated naturally 140
Bones Vitamin A necessary for health of 47; Boron maintains health of 90; healthy bone will bend rather than break 92
Bones brittle excess Fluoride 104; lack of Calcium 94; Vitamin D and Calcium reduced hip fractures 43% 78
Bones malformed lack of Calcium 94; lack of Vitamin D 78; wrong Phosphorus-Calcium balance 119; successfully prevented naturally 141
Bones painful lack of Vitamin D 79
Bones soft lack of Vitamin D 79
Bones slow fracture healing lack of Vitamin C 75
Bones weak lack of Magnesium 112; lack of Vitamin C 47; wrong Phosphorus-Calcium balance 119
Boron section on **90**; food chart 91
Bowel cancer adequate Folate reduces incidence 75% 62
Bran raw with meals can bind Zinc and other minerals 134

Bread 10 minute wholemeal recipe 165; shop bought wholemeal mostly half white flour 14
Breast feeding higher RDI requirements 11
Breast feeding insufficient milk lack of Riboflavin 52; lack of Vitamin C 76
Breast lumps See **Fibrocystic Breast**
Breathing deep important to avoid constipation 38; slowing can calm stress 85; therapy known as rebirthing 153
Bronchitis successfully treated naturally 140
Bruising easy lack of Vitamin B12 66; lack of Vitamin C 76; lack of Vitamin K 88
Bulimia lack of Zinc 136

Cadmium Bluff oysters high in 53; excess agravates stress 26; NZ soils contaminated by Naru fertiliser 53; excess can raise cholesterol levels 25; Vitamin C helps expell 74; Zinc helps expell 136
Caffeine can deplete Inisitol levels 69; excess agravates stress 26; excess can raise cholesterol levels 24
Calcium Aluminium inhibits absorbtion 89; Boron helps absorbtion 90; excess Sodium can deplete 129; **section on 92**; food chart 95; mixes with cholesterol to narrow and harden arteries 23; needs Vitamin D for good absorbtion 78; one part Phosphorus to two parts Calcium critical 119; raw bran can hinder absorbtion of 37; Vitamin K helps keep in bones 88; with Vitamin D markedly reduced hip fractures 78
Calcium growths lack of Vitamin B12 66
Calories 1 kg fat contains 9000 31; to burn off chocolate bar (chart) 18; controlling intake of 153; daily needs (chart) 16; excess major cause of ill health 40; **section on 16**; food chart 19; NZ survey results 17; percent in high fat foods (chart) 32; excess can raise cholesterol levels 25; usage chart 17
Cancer brisk walking cuts risk 20% 80; dramatic fall-off at close of World War 2 16; eleven times lower when Selenium and Vitamin E levels high 124; 'terminally ill' patients went home healed 142; lack of Vitamin A 47; lack of Selenium 126; lack of Zinc 136; radical surgery does more harm than good 148; untreated victims tend to live up to four times longer 148
Cancer bowel fibre does not protect from 37; adequate Folate reduces incidence 75% 62
Cancer cervical lack of Vitamin C 76
Cancer lung lack of Folate in smokers 63; lack of Vitamin B12 in smokers 66
Cancer oesophagus lack of Molybdenum a factor 117
Cancer prostate lack of Vitamin E 82; lack of Boron 90; lack of Selenium 124; lack of Zinc 136
Cancer skin lack of Calcium 94; lack of Selenium 126; Vitamin B6 cream can regress Melanoma 58
Cancer stomach lack of Vitamin C 76
Cancer throat lack of Vitamin C 76

Candida excess sugar 45; lack of Iron 109
Canola oil contains useful amounts Omega 3 33
Carbohydrates athletes perform better on 21; compared to honey (chart) 21; **section on 21;** RDI of 22; comparison with high fat foods 30
Carotene See **Vitamin A**
Carpal Tunnel Syndrome See **Repetitive Strain Injury**
Cataract See **Eyes cataract**
CFS See **Chronic Fatigue Syndrome**
Chelated minerals bound with substance to aid absorption 132
Chlorine how to rid water of smell 97; section on 96
Cholesterol section on 23; blood level guidelines 28; fibre can bind excess 37; vegetarian societies have lowest rates in world 26
Cholesterol high section on 23; excess Cadmium 25; excess alcohol 24; excess caffeine 24; excess sugar 24; lack of Chromium 99; lack of Copper 101; lack of fibre 38; lack of Iodine 106; lack of Magnesium 112; lack of Manganese 115; lack of Niacin 56; lack of unsaturated fat 34; lack of Vitamin C 75; lack of Zinc 136; long term stress 24; excess calories 24; smoker twelve times more likely to have 29
Cholesterol high density good and essential for body 23; Lecithin believed to maintain optimum level 68; Niacin can increase 35% 55; excess Zinc can lower levels of 136
Cholesterol low density harmful and can clog arteries 23; Niacin can lower 30% 55; Vitamin C can lower 10% 74
Cholesterol lowering drugs hinder manufacture of Vitamin D 78
Chromium section on **98;** protects against high cholesterol 24
Chronic Fatigue Syndrome lack of Sodium a factor 129; linked with low blood pressure 113
Cleft palate lack of B6 during pregnancy 58; lack of Folate during pregnancy 62
Cobalt part of Vitamin B12 65; some NZ soils seriously low in 65; excess can raise cholesterol levels 25
Cod liver oil high in Omega 3 31
Coffee excess can drop Vitamin B6 levels 31; excess can lower Vitamin B6 levels 58; excess can raise blood fat level 31; hinders Iron absorption 39% 108
Cold feeling the exercise raises body metabolism rate 155; lack of Iodine 106
Cold hands and feet lack of Folate 63; sign of hypoglycaemia 44
Colds can avoid with 3000 mg Vitamin C daily 74; complete absence on restricted rations 16; lack of Oxygen 42; lack of Pantothenic Acid 71; lack of Vitamin C 75; I rarely get 149
Colloidal minerals able to be permanently suspended in water 132, 137; seawater minerals and elements colloidal 107

Colliidal Silver natural antibiotic, how to make 137
Colourings food harmful additives listed with * alongside 159
Concentration poor lack of Boron 90
Confusion lack of Magnesium 112; lack of Thiamine 50; lack of Vitamin B12 65
Constipation excess Iron 109; how to avoid 38; lack of fibre 37; lack of Potassium 122; lack of Sodium 129; lack of Sulphur 131; lack of Thiamine 50
Convulsions in infants whooping cough vaccine 148; mother lacking Vitamin B6 59; lack of Magnesium 112; lack of Manganese 115
Co-ordination lack of lack of Boron 90; lack of Magnesium 112; lack of Vitamin B6 66
Copper bracelets can reduce arthritic pain 102; deficiency can be due to excess Zinc 136; excess agravates stress 26; **section on 100;** food chart 102; Molybdenum helps absorb 117; excess can raise cholesterol levels 25
CoQ10 lack of can raise cholesterol 25; **section on 84;** food chart 85
Cortisone hinders manufacture of Vitamin D 78
Cot death DPT (diphtheria) immunisations 148; lack of Selenium 126; lack or excess of Thiamine 49
Cramp leg lack of Calcium 94; lack of Magnesium 112; lack of Pantothenic Acid 71; lack of Potassium 122; lack of Sodium 129; lack of Vitamin E 82
Cystic Fibrosis lack of Selenium 126

Dandruff excessive lack of Selenium 126; lack of Zinc 136
Daniel in Babylon test us for ten days 41
Death major causes of 146
Deformities See **Malformities**
Depression excess Lecithin 68; lack of Magnesium 112; lack of Pantothenic Acid 71; lack of Potassium 122; lack of Thiamine 50; lack of Vitamin B6 59; lack of Vitamin C 75; lack of Zinc 136; sign of hypoglycaemia 44
Diabetes dramatic fall-off at close of World War 2 16; excess calories can cause 16; excess dietary fat main cause of 31; hypoglycaemia can lead to 44; lack of Chromium a factor 99; lack of Manganese 115; lack of Zinc 136; often found with heart disease 44; serious disorder doubling every 15 years 31; successfully treated naturally 140
Diarrhoea can occur when changing diet 37; common following antibiotics 87; excess Magnesium 112; excess Vitamin C 76
Digestion poor lack of Iron 109; lack of Vitamin C 76
Digestive system complex and mysterious 138; fibre necessary for proper functioning 38; three simple rules regarding 138
Dirt eating lack of Zinc 136
Diverticulitis constipation can lead to 37; lack

of fibre 38; successfully treated naturally 140
Dizziness lack of Sodium 129
Dolomite raw form of Calcium and Magnesium 94; should have Vitamin D added 94
Dreams no recall, lack of Vitamin B6 59; Vitamin B6 can affect 58
Drug addiction hypoglycaemia common factor 44; lack of Folate 63; lack of Niacin 56; Vitamin C helps overcome 75; successfully treated naturally 140
Drugs pharmaceutical mostly alleviate symptoms only 145
Dwarfism lack of Zinc 136; successfully treated naturally 140

Ear noises in lack of Manganese 115
Easy bruising lack of Vitamin C 76
Eczema lack of Riboflavin 53; successfully treated naturally 140
Edema can cause a weight problem 151; excess Sodium 129; lack of Potassium 122; lack of protein 41; lack of Vitamin B6 59
Eggs egg-nog every nutrient for human life 144; quality source of Sulphur 131; raised cholesterol only 2% in non-smokers 25; rich source of nutrients 54
Encephalitis measles shots 148
Epilepsy lack of Magnesium 112; can improve on Folate and B12 supplements 62; successfully treated naturally 141
Epsom salts (Magnesium Sulphate) used as laxative 111
Eskimos consume huge amounts fat 30; seldom develop clogged arteries 24; women suffer from osteoporosis 92
Exercise antidote to stress 72; how to beat boredom of 156; can improve health dramatically 145; for less vigorous types 152; for vigorous types 152; lack of a major cause of disease 30; thirty minutes daily of benefit 116; personal testimony of 150; raises metabolism rate 155; rapid weight loss method of 155
Eyes bloodshot lack of Riboflavin 52
Eyes blurred vision sign of hypoglycaemia 44
Eyes cataracts lack of Calcium 94; lack of Riboflavin 52; lack of Selenium 126; lack of Vitamin A 47; lack of Vitamin B6 59; lack of Vitamin C 75; lack of Zinc 136; Riboflavin reported to cure 52; in infants lack of Vitamin D 79
Eyes dark skin under lack of Riboflavin 52
Eyes dry or rough lack of Vitamin A 47
Eyes glaucoma lack of Vitamin C 75; successfully treated naturally 141
Eyes Irritable lack of Riboflavin 52
Eyes macular degeneration lack of Iodine 107; successfully treated naturally 141
Eyes over-sensitive light lack of Riboflavin 52
Eyes poor night vision lack of Vitamin A 47; lack of Zinc 136
Eyes protruding lack of Iodine 106
Eyes red eyelids lack of Niacin 56; lack of Riboflavin 52; lack of Vitamin B6 59
Eyes twitchy lids lack of Vitamin B6 59
Eye sight blindness in elderly lack of Zinc 136; lack of Vitamin A 47; result of diabetes 34
Eye sight weak in elderly lack of Folate 63; lack of Vitamin B12 66; lack of Zinc 136
Face haggard lack of Vitamin C 76
Face swollen See also **Edema**; lack of Vitamin B12
Fainting sign of hypoglycaemia 44
Fasting health benefits of 157
Fatigue See **Low energy**
Fat 37% of total calories in NZ 31; high fat foods (chart) 32; consumption not increased significantly for 90 years 27; RDI of 22; **section on** 30; one kg contains 9000 calories 31; percent in low fat foods (chart) 32; thicken blood during digestion 24
Fat Omega 3 type of unsaturated fat with blood thinning qualities 31; one of three main fatty acids 33; protects against blood clotting 33; Cod liver oil,Sardines and Flax oil high in 31; Soya Bean, Canola oil contain useful amounts 33
Fat Omega 6 one of three main fatty acids 33
Fat Omega 9 one of three main fatty acids 33
Fat polyunsaturated not significantly different from unsaturated fat 33
Fat saturated excess can raise cholesterol 26; **section on** 33; food chart 36; recommended 2 to 1 in favour of unsaturated 34; NZ diet currently 1 to 1 34
Fat unsaturated lack of can raise cholesterol 26; **section on** 33; food chart 36; recommended 2 to 1 in favour 34; NZ diet currently 1 to 1 34
Fatty acids See **Fat saturated** , **Fat unsaturated**
Fearfulness in children lack of Iron 109
Feet burning sensation in lack of Pantothenic Acid 71
Feet cold lack of Folate 63; sign of hypoglycaemia 44
Feet numb and tingling lack of Thiamine 50
Fertilisers Finland adds Selenium to 124; NZ soils contaminated by Cadmium 53; Zinc not returned to soil by NPK 134
Fibre lack of can raise cholesterol 25; **section on** 37; food chart 39
Fibrocystic Breast lack of Iodine 107; successfully treated naturally 141
Fingers lumps on lack of Vitamin B6 59
Fingers numb or tingly excess Fluoride 104; lack of Vitamin B12 66
Fingers stiff swollen joints lack of Vitamin B6 59
Fish 225 gms a week lowers risk of stroke 24
Fish oil can lower cholesterol 24; tends to be high in Omega 3 30; Vitamin D found in 78
Fitness activity level necessary for 152; resting pulse guide to (chart) 150; will normally maintain optimum weight 152

Fits lack of Vitamin C during pregnancy 75
Flax oil high in Omega 3 31
Flour white 75% less fibre than wholemeal 38; one of three deadly white processed foods 128; has typically lost 60% to 85% of Vitamins and minerals 145
Flu can be more virulent if lacking Selenium 124; now enjoy total freedom from 149; fresh air necessary to avoid 42
Fluid retention See **Edema**
Fluoride odourless tasteless in drinking water 96; **section on 103**; food chart 104
Folate 90% of alcoholics deficient 15; alcohol can hinder absorbtion 15; **section on** 62; food chart 64; lack of agravates stress 26; need for doubles in pregnancy 64; similar in action to B12 62; reduces incidence of bowel cancer 75% 62
Folic acid See **Folate**
Food additive code numbers section on **159**
Food tables explanation of 13
Free radicals can form in unsaturated fats when heated 81

Gall bladder cholesterol can clog 23
Gallstones excess sugar in diet 45; lack of fibre 38; lack of unsaturated fat 34
Gangrene can result from diabetes 34
Garlic natural blood thinner 35
Germs not first cause of disease 83
Glaucoma See **Eyes glaucoma**
Glucose muscles and brain use for fuel 43; sugars converted to body 43
Glycaemic index ranking of carbohydrates 22
Goitre enlarged thyroid gland due to lack of Iodine 106; excess Cobalt 66
Gout lack of Folate 63; lack of Molybdenum 117; successfully treated naturally 141; wrong Phosphorus-Calcium balance 119
Growing pains lack of Vitamin E 82
Growth slow lack of fat in diet 34; lack of Iron 109; lack of protein 41; lack of Riboflavin 52; lack of Thiamine 50; lack of Zinc 136
Growth stunted lack of Selenium 126; lack of Vitamin A 47; successfully treated naturally 141; wrong Phosphorus-Calcium balance 119
Gum disease lack of CoQ10 84; lack of Niacin 56; wrong Phosphorus-Calcium balance 119
Gums abnormally red lack of Vitamin C 75
Gums bleeding lack of Vitamin C 76

Haemorrhoids constipation often leads to 37; how to avoid 38; lack of fibre 38
Hair Sulphur necessary for gloss and curliness 131
Hair dull lack of fat in diet 34; lack of protein 41; lack of Vitamin A 47; lack of Sulphur 131
Hair lack of pigmentation lack of Copper 101
Hair loss excess Vitamin A 47; lack of Biotin 86; lack of Protein 41; lack of Riboflavin 53; lack of Zinc 137

Hair prematurely grey lack of Biotin 86; lack of Folate 63; lack of Vitamin B12 66; lack of Sulphur 131; lack of Zinc 136
Hair stiff lack of Biotin 86; lack of Zinc 136
Hands numb or tingling lack of Thiamine 50; lack of Vitamin B6 59
Hands trembling lack of Magnesium 112
Hay fever excess sugar can increase sensitivity to 45; lack of unsaturated fat a factor 34; successfully treated naturally 141; I enjoy total freedom from 149
Headache sign of hypoglycaemia 44; excess Fluoride 104; lack of Potassium 122; how to avoid 87
Headache migraine lack of Niacin 56
Healing advice from Dr Ulric Williams 149; cause of problem must be rectified 147
Health factors that play havoc with 145; first law of 18
Hearing loss lack of Niacin 56; lack of Vitamin B12 66
Heart attack section on Artery Disease **23**; lack of Chromium 99; lack of CoQ10 84; lack of Folate 62; lack of Iodine 106; lack of Magnesium 112; lack of Manganese 115; lack of Niacin 56; lack of Selenium 126; lack of unsaturated fat 34; lack of Vitamin C 75; lack of Vitamin D 79; lack of Vitamin E 82; lack of Zinc 136; nearly 50% higher during winter 78; prolonged stress can cause 26; symptoms of 29; usually caused by blood clot 24
Heart beat resting pulse guide to fitness (chart) 150
Heart beat irregular lack of Calcium 94; lack of Magnesium 112; lack of Potassium 122; lack of Thiamine 50; successfully treated naturally 141
Heart beat rapid lack of Thiamine 50
Heart disease section on Artery Disease **23**; 50% increase when jobs were at risk (study) 26; Australia 20% lower rate than NZ 124; excess sugar 44; excess calories 16; lack of Lecithin 68; lower when water high in Magnesium 111; often found with diabetes 44; risk drops to normal after quitting smoking ten years 29
Heart muscle can be damaged by flu if lacking Selenium 124
Heart weak lack of Iron 109
Hepatitis lack of Lecithin 68
Herbalism similar success rate to orthodox medicine 142
High blood pressure See **Blood pressure high**
Hip fractures Vitamin D and Calcium reduce by 43% 78
Homeopathy similar success rate to orthodox medicine 142
Honey compared to carbohydrate (chart) 21
Hostility raises cholesterol 23
Hyperactivity 59; excess Aluminium 89; excess Copper 101; food allergies major cause 58; lack of Magnesium 112; lack of unsaturated fat 34; lack of Zinc 136

Hypoglycaemia wildly fluctuating blood sugar levels 44; 65% incidence among schizophrenics 44; 95% incidence among alcoholics 44; common in hyperactive (ADD) children 44; lack of Chromium 99; lack of Manganese 115; lack of Zinc 136; often forerunner of diabetes 44; root cause of most anti-social behaviour 44; symptoms of 44

Immunity low lack of Vitamin A 47; lack of CoQ10 84; lack of Oxygen 42; lack of Iron 109; lack of Zinc 136; high blood sugar 45

Impotency lack of Zinc 136

Indigestion lack of Pantothenic Acid 71

Infections repeated lack of Pantothenic Acid 71; lack of Selenium 126; lack of Vitamin C 75; lack of Vitamin E 82; lack of Zinc 135; Colloidal Silver proven effective 137

Infections respiratory dramatic fall-off at close of World War 2 16; excess calories 16

Infertility lack of Vitamin C 76; lack of Selenium 126; lack of Zinc 136

Inositol lack can agravate stress 26; lack can raise cholesterol 25; **section on 69**; food chart 70

Insect bites large doses of Thiamine can repel 49

Insomnia dramatic fall-off at close of World War 2 16; excess calories can cause 16; excess Copper 101; excess Sodium 129; lack of Calcium 94; lack of Inisitol 69; lack of Manganese 115; lack of Niacin 55; lack of Pantothenic Acid 71; lack of Thiamine 50; lack of Vitamin B6 59; lack of Vitamin C 75; lack of Zinc 136

Iodine excess Cobalt can cause deficiency 66; lack of can raise cholesterol 25; **section on 106**; food chart 107

Iron Zinc protects against excess 110; Copper helps absorb 101; **section on 108**; food chart 110; Molybdenum helps us absorb 117; Vitamin B6 necessary for absorbtion 58; Vitamin C protects against excess 74; how to absorb more 110; Pawpaws and pauas a rich source 108

Irritability lack of Calcium 94; lack of Folate 63; lack of Vitamin B6 59; sign of hypoglycaemia 44

Joint disorders lack of Vitamin B12 66
Joints aching lack of Sodium 129
Joints painful lack of Vitamin B12 66; lack of Vitamin C 76

Kelp powder source of sea minerals 107
Kidney failure excess Aluminium 89; excess protein 41; lack of Vitamin C during pregnancy 75; lack of Zinc 136
Kidney stones 112; excess sugar 45; lack of Calcium 94; successfully treated naturally 141

Lead levels in NZ foods dropped 53; Vitamin C assists in expelling 74
Lead poisoning symptoms similar to lack of Boron 90; successfully treated naturally 141
Learning difficulties See **Attention disorder, Hyperactivity**
Lecithin can be manufactured by healthy liver 68; section on **68**
Legs aching lack of Vitamin D 79
Legs bowed lack of Vitamin D 78
Leg cramps See **Cramp leg**
Legs numb or tingly excess Fluoride 104; lack of Calcium 94; lack of Vitamin B12 66; lack of Thiamine 50
Legs painful muscles lack of Thiamine 50
Legs swelling of See **Edema**
Legs trembling lack of Calcium 94; lack of Magnesium 112; lack of Niacin 56; lack of Vitamin B6 59; sign of hypoglycaemia 44
Lipids See **Fat**
Lips cracked lack of Riboflavin 52
Lips red burning lack of Riboflavin 53
Listlessness lack of Vitamin B12 65; See also **Apathy, Low energy**
Liver loses ability to make CoQ10 as we age 84; body's storehouse of nutrients 46; stores and combines protein 40
Liver disorders excess Aluminium 89; lack of Vitamin K 88; lack of Zinc 136
Longevity cheerful people live 19% longer 56; five laws of 122; life expectancy in Japan surpasses NZ by four years 128; reduced if lacking CoQ10 84; mice treated with CoQ10 lived 56% longer 84; vegetarian societies have longest life spans 26
Low appetite lack of Vitamin B6 56
Low blood pressure See **Blood pressure low**
Low energy lack of Copper 101; lack of CoQ10 84; lack of Iodine 106; lack of Iron 109; lack of Magnesium 112; lack of Niacin 55; lack of Pantothenic Acid 71; lack of Potassium 122; lack of Sodium 129; lack of Thiamine 50; lack of Vitamin B12 66; lack of Vitamin B6 59; lack of Vitamin E 82; lack of Zinc 136; sudden fatigue sign of hypoglycaemia 44; wrong Phosphorus-Calcium balance 119
Low sperm count lack of Selenium 126; lack of Zinc 136
Lung infection See **Cystic Fibrosis**

Magnesium alcohol can hinder absorbtion 15; Boron helps absorbtion 90; lack can cause 'shakes' of alcoholic 15; lack can raise cholesterol 25; **section on 111**; food chart 113; raw bran can hinder absorbtion of 37; soft water does not normally contain 25
Malformities lack of Folate during pregnancy 62; lack of Iodine in mother 106
Manganese lack of can raise cholesterol 25; section on **115**; food chart 116
Maniac-depressive disorder lack of Lecithin 68
Margarine similar to animal fat 24; has characteristics of saturated fat 33
ME See **Chronic Fatigue Syndrome**
Melanoma See **Cancer skin**
Memory poor short term lack of Boron 90; lack of Lecithin 68; lack of Thiamine 50; lack of Vitamin B12 65; successfully treated naturally 141

Menstrual tension See **PMT**

Mental alertness reported among those with high intake of Vitamin C 74

Mental disorders 50% incidence of hypoglycaemia among mental patients 44; excess Copper 101; lack of Calcium 94; lack of Folate 63; lack of Inisitol 69; lack of Manganese 115; lack of Niacin 55; lack of Thiamine 50; lack of Vitamin C 75; lack of Zinc 136; Vitamin C widely lacking in those suffering from 75

Mental retardation lack of Folate during pregnancy 62

Mercury Vitamin C assists body in expelling 74

Metals toxic See **Toxins in body**

Migraine See **Headache migraine**

Milk can supply Vitamin B12 needs 67; difficult to digest for many Maori and Pacific Islanders 93; fed them only apples and milk 142; warm milk can promote sleep 69; homogenised condemned as harmful 144; 20% more solids in low/non-fat 14; rich natural source of Riboflavin 53; still richest source of Calcium 93; lacking in Vitamin C 74; yoghurt more easily digested 93

Miscarriage lack of Folate during pregnancy 62; successfully prevented naturally 141

Molasses rich in trace minerals 99

Molybdenum section on 117; food chart 118

Mongolism lack of Zinc during pregnancy 136; successfully prevented naturally 141

Mormon health impressive statistics on 157; Word of Wisdom revelation 158

Mouth cracks/sores at corners lack of Riboflavin 52; lack of Vitamin B6 59; lack of Iron 109

Mouth malformities lack of Vitamin B6 during pregnancy 59

Mouth ulcers excess sugar in diet 45

Multiple Sclerosis lack of Manganese 115; successfully treated naturally 141

Muscle building can produce spectacular results in 3 years 156

Muscle cramp See **Cramp**

Muscle pain excess Aluminium 89; lack of Vitamin C 76

Muscle wasting lack of protein 41

Muscles aching lack of Sodium 129

Muscles numb excess Aluminium 89

Muscles poor control of See **Co-ordination lack of**

Muscles poorly developed lack of Vitamin D 79

Muscles sagging lack of Copper 101; lack of Potassium 122

Muscles weak excess Aluminium 89; lack of CoQ10 84; lack of Iron 109; lack of Vitamin B6 66; lack of Vitamin D 79; wrong Phosphorus-Calcium balance 119

Nails brittle excess Vitamin A 47; lack of Zinc 136

Nails concave lack of Iron 109

Nails white spots lack of Zinc 136

Naturopathy similar success rate to orthodox medicine 142

Nausea lack of Biotin 86; lack of Vitamin B6 59

Negative thinking how to break habit of 51; major cause of ill health 51; worst cause of disease - Dr Williams 149

Nerve disorders lack of Copper 101

Nerve pains lack of Sodium 129

Nervous twitches lack of Calcium 94; lack of Lecithin 68

Nervousness lack of Magnesium 112; lack of Manganese 115; lack of Thiamine 50; lack of Vitamin C 75; wrong Phosphorus-Calcium balance 119

Niacin 60% of alcoholics benefited from 15; lack can cause 'shakes' of alcoholic 15; lack of agravates stress 26; lack can raise cholesterol 25; **section on** 55; food chart 57; Phosphorus-Calcium balance critical to absorption of 119

Niacinamide form of Niacin 55

Nicotinic Acid See **Niacin**

Nose bleeds lack of Vitamin K 88

Numbness lack of Calcium 94; lack of Magnesium 112; See also **Legs numb or tingly, Fingers numb or tingly, Muscles numb**

NZ nutrition survey calorie intake 17; carbohydrate, sugar, protein, fat intake, 22; pesticide contamination low 53; most Vitamins and minerals supplied in sufficient amounts 138; weight statistics 151

Obesity lack of Iodine 106; when unfit increases risk of artery disease 60% 25; section on slimming 151

Oestrogen lack of Boron doubled production of 90

Oils See **Fat**

Old age spots lack of Selenium 126

Omega 3 See **Fats Omega 3**

OOS See **Repetitive Strain Injury**

Orthodox medical treatment advances impressive but some aspects profit driven 145; tend to treat symptoms rather than causes 147; we must accept most of blame 147

Osteoarthritis See **Arthritis**

Osteopathy similar success rate to orthodox medicine 142

Osteoporosis 1200 mg of Calcium daily can prevent 93; common among Eskimo women on high protein diets 92; excess Aluminium 89; excess protein 41; fit and active generally do not develop 92; lack of Boron 91; lack of Calcium 94; lack of Magnesium 112; lack of Vitamin D 79; lack of Vitamin K 88; linked with high protein intake 92; oestrogen appears to protect against 92; vegetarians seldom develop 92; wrong Phosphorus-Calcium balance 119

Oxygen essential for immune system 42; most important nutrient of all 42; helps avoid colds and flu 42

Paua rich source of Iron 108
Pale skin See **Anaemia**
Palpitations See **Heart beat irregular**
Pancreas can only take so much abuse 44; seriously affected by lack of Chromium 98
Panic attacks lack of Thiamine 50; sign of hypoglycaemia 44
Pantothenic Acid alcohol can hinder absorbtion 15; section on **71**; food chart 73
Paranoia sign of hypoglycaemia 44; successfully treated naturally 141
Parkinson's Disease some cases cured by injecting Vitamin B6 58; lack of Vitamin B6 59
Pawpaw rich source of Iron 108
Peace inner methods to attain 152
Peas green close to perfect natural food 144
Pectin limits harmful cholesterol 27
Pesticides low in NZ foods 53
Phosphorus section on **119**; food chart 120
Piles See **Haemorrhoids**
PMT lack of Calcium 94; lack of Magnesium 112; lack of Vitamin B6 59; lack of Vitamin E 82; successfully treated naturally 141
Poisons Vitamin C helps expel 75; Zinc and Vitamin C can expel 137
Polio most cases now caused by vaccine 148
Pollutants low resistance to lack of Selenium 126
Pollutants should not trouble well nourished person 139
Post-natal depression erratic sleep pattern or excess Copper 100; lack of Zinc 136
Potassium excess Sodium can deplete 129; increase intake when sweating heavily 130; section on **121**; food chart 123
Pre-eclampsia lack of Vitamin C 75
Pregnancy coming off the pill 60; babies heavier due to Zinc supplementation 135; can halve body levels of Chromium 98; erratic sleep can cause mental instability after 100; Folate, Vitamin C, Iron, Zinc should be doubled 11; higher RDI requirements 11; lack of Vitamin B6 can cause toxaemia 59; Magnesium can relieve painful contractions 111; need for Folate during 64; stretch marks due to lack of Zinc 136; Vitamin C helps prevent pre-eclampsia 75
Pregnancy nausea lack of Vitamin K 88
Premature birth lack of Folate 63
Pre-menstrual tension or PMS See **PMT**
Pritikin program to heal heart disease 31
Prostate cancer See **Cancer prostate**
Prostate enlarged lack of Magnesium 112; lack of unsaturated fat 34; lack of Zinc 136; successfully treated naturally 141
Protein excess hinders absorbtion of Calcium 92; section on **40**; food chart 42
Psychiatric patients commonly lack Folate 62; widely lacking in Vitamin C 75
Pulse See **Heart beat**

Pulse test can identify an allergenic food 140
Radiation Iodine protects against 106; Pantothenic Acid protects against 71
Rapid weight loss method lose 3 kg week 154
RDI Australian last revised 1991 11; explanation of 11
Relaxation words can markedly affect our 73
Repetitive Strain Injury lack of Magnesium 112; lack of Vitamin B6 59
Resentment can only exist with blame 35; major cause of ill health 35
Retinol See **Vitamin A**
Riboflavin does not discolour urine when taken naturally in milk 143; in pill form immediately expelled in urine 143; **section on 52**; food chart 54
Rice white poor provider of Protein 40; typically lost 60% to 85% of vitamins and minerals 145
Rickets lack of Vitamin D 79
RSI See **Repetitive Strain Injury**
Salt refined one of three deadly white processed foods 128; had most of 77 minerals and elements removed 107; See also **Sodium**
Salt rock sea salt mined from ancient sea beds 107
Salt unrefined sea all minerals required by body found in 107
Schizophrenics 65% incidence of hypoglycaemia among 44; can become normal after Calcium injections 94; lack of Niacin 55; lack of Vitamin B12 66; nearly all lack Vitamin C 75; commonly respond to treatment with Folate, Niacin, and Zinc 62; successfully treated naturally 141
Scurvy lack of Vitamin C 76
Sea water contains 77 colloidal minerals and elements 107; health benefits of 107; can rapidly heal acne and other skin infections 107
Sedatives See **Tranquillisers**
Selenium good sources of 127; lack of can raise cholesterol 25; linked with longevity 122; protects against high cholesterol 24; **section on 124**; food chart 127; required to manufacture CoQ10 84; works hand in hand with Vitamin E 81
Senility excess Aluminium 89; lack of Folate 62; lack of Iodine 107; lack of Vitamin B12 66
Sexual abuse overcoming trauma of 153
Sexual development slow lack of Zinc 136
SID See **Cot death**
Skin brown old age spots lack of Selenium 126
Skin cancer See **Cancer skin**
Skin coarse lack of Vitamin A 47
Skin darkened on face/hands lack of Niacin 56
Skin dry excess Vitamin A 46; lack of Biotin 86; lack of fat in diet 34; lack of Iodine 107; lack of Potassium 122; lack of Sulphur 131; lack of Vitamin C 76; successfully treated naturally 140
Skin early wrinkling lack of Copper 101; lack of Zinc 136

173

Skin flaking excess Vitamin A 46
Skin greyish hue lack of Biotin 86; lack of Iron 109
Skin infections sea water can rapidly heal 107; ; Colloidal Silver proven effective 137
Skin itchy excess Vitamin A 47; lack of Riboflavin 53
Skin oily lack of Vitamin A 47
Skin pale See **Anaemia**
Skin patchy loss of pigmentation lack of Copper 101; lack of Vitamin B12 66
Skin pin head spots lack of Vitamin C 76
Skin purplish lack of Vitamin K 88
Skin raised hair follicles lack of Vitamin C 76
Skin rash around nose, mouth lack of Biotin 86
Skin rashes excess Vitamin C 76
Skin red greasy on face lack of Riboflavin 52
Skin scaly lack of Biotin 86; lack of Vitamin C 76
Skin yellow excess Vitamin A 47
Sleep three ways to a good 69
Sleeping pills hinder body manufacturing Vitamin D 78; how to break addiction to 114
Slow growth See **Growth slow**
Slow learning See **Attention disorder**
Smell poor sense of lack of Vitamin A 47; lack of Zinc 136
Smoking raises cholesterol levels 24; destroys 27% of Vitamin C in blood 75; hinders absorption of Zinc, destroys Vitamin C 25; lack of Niacin can increase addiction to 55; smokers twelve times more likely to suffer high cholesterol 29; tobacco not for the body 158
Sodium increase intake if sweating heavily 130; should not increase blood pressure if Potassium intake increased 121; **section on 128**; food chart 130
Soya Bean oil contains useful amounts of Omega 3 33
Spina Bifida lack of Folate during pregnancy 62
Spinal disorders lack of Manganese 115
Spine malformities lack of Vitamin D during pregnancy 79
Stillbirth lack of Vitamin B6 during pregnancy 59
Stomach acid lacking lack of Iron 109
Stomach rumbling sign of anxiety 139
Stomach ulcers lack of Folate 63
Stress 50% increase in heart disease when jobs at risk 26; agravated by excess caffeine, Copper, Cadmium 26; agravated by lack of Niacin, Inositol, Folate 26; antidote to 72; can cause back pain 85; can drop Vitamin A levels dramatically 46; can lower body levels of Vitamin B6 58; can raise cholesterol levels 24; how to lower 35; increased enormously due to breakdown of family, uncertain employment 27; list of long term emotional stresses to avoid 144; prolonged stress cause of high blood pressure 26; raised cholesterol of shift workers study 25; role in heart disease 112; three rules to minimise 61

Stretch marks lack of Zinc 136
Stroke section on Artery Disease **23** excess Sodium (factor) 129; how to lower risk 40% 123; lack of Chromium 99; lack of Copper 101; lack of CoQ10 84; lack of Iodine 106; lack of Magnesium 112; lack of Manganese 115; lack of Niacin 56; lack of Potassium 122; lack of Selenium 126; lack of unsaturated fat 34; lack of Vitamin C 75; lack of Vitamin E 82; lack of Zinc 136; prolonged stress can cause 26
Sugar excess can lower B12 66; excess can raise cholesterol levels 24; levels in processed foods 45; **section on 43**; obesity and sugary drinks linked 153; RDI of 22; body requires Chromium to process 98; tame that craving 45
Sugar white one of three deadly white processed foods 128
Sulphur section on **131**; food chart 133
Summary six key factors of health 144
Sunburn Calcium protects against 94
Supplements B family Vitamins better absorbed in milk powder preparations 143; can they help us? 142; major defects in diet cannot be corrected with 143; maybe required to rectify long standing deficiencies 144; some unable to be utilised by body 143
Surgery analogy of car engine 147; radical surgery on cancer can do more harm than good 148
Swallowing painful lack of Iron 109
Swelling of tissue See **Edema**

Taste poor sense of lack of Zinc 136
Tea hinders Iron absorption 64% 108
Teeth Molybdenum assists in forming enamel 117; tooth enamel largely Magnesium 111; Vitamin A regulates spacing, straightness of 47
Teeth brittle excess Fluoride 104
Teeth crooked lack of Vitamin A 47
Teeth malformed wrong Phosphorus-Calcium balance 119
Teeth mottled excess Fluoride 104
Teeth weak lack of Vitamin A 47; wrong Phosphorus-Calcium balance 119
Thiamine section on **49**; food chart 51
Thrush excess sugar in diet 45
Thyroid swollen See **Goitre**
Tingling lack of Magnesium 112
Tiredness lack of Magnesium 112; See also **Low energy**
Tongue dark red lack of Vitamin B12 66
Tongue red burning lack of Niacin 56; lack of Riboflavin 53
Tongue sore red lack of Folate 63; lack of Vitamin B6 59
Tongue ulcers excess sugar in diet 45
Tonsil infection cause of bad breath 80
Tooth decay can allow entry into bloodstream of dangerous bacteria 104; how to avoid 105; lack of Magnesium 112; lack of Molybdenum

117; lack of Phosphorus 119; lack of Vitamin C 76; lack of Vitamin D 79; nutrients that play role in preventing 104; sticky sweets and chocolate major offenders 44; Vitamin A protects against 47

Toxaemia lack of Folate during pregnancy 62; lack of Vitamin B6 59; successfully treated naturally 141

Toxins in body lack of Vitamin C 75; Zinc and Vitamin C can expel 137

Tranquillisers lack of Niacin can contribute to dependence on 55; how to break addiction to 114; Niacin can take place of 55; Calcium a natural 94

Trembling lack of Calcium 94; lack of Magnesium 112; lack of Niacin 56; lack of Vitamin B6 59; sign of hypoglycaemia 44

Tuberculosis lack of Vitamin D increases risk 79; lack of Vitamin A 47

Twitches See **Nervous twitches**

Ulcers successfully treated naturally 141; See also **Infections repeated**

Uric acid Folate helps control level of 63

Urinary irritation excess Vitamin C 76

Urine Riboflavin supplement can turn bright yellow 52

Vaccines words of Dr Mendelsohn 148

Varicose veins excess sitting 38; how to minimise 64; lack of Copper 101

Vegetable oil can lower cholesterol 24

Vegetarians lowest cholesterol rates in world 26; deficiencies of B12 found among 65; milk can supply B12 needs 67; develop clogged arteries under stress 25; retain higher proportion of Calcium 93

Violent behaviour levels of Cobalt very low in violent offenders 65; hypoglycaemia often a cause 44; close connection with severe mineral imbalances 137

Vitamin A important anti-oxidant 46; section on 46; food chart 48

Vitamin B1 See **Thiamine**

Vitamin B2 See **Riboflavin**

Vitamin B3 See **Niacin**

Vitamin B6 can regress Melanoma if applied as a cream 58; excess coffee can drop body levels 31; **section on 58**; food chart 60; test for deficiency 58

Vitamin B12 Vitamin B6 helps absorbtion 58; section on **65**; food chart 67

Vitamin C assists body in ridding itself of toxic metals 74; can double Iron absorbtion 108; disorders that respond to 77; double intake during pregnancy 11; increased mental alertness reported with high intake 74; lack of can raise cholesterol 25; lack of Vitamin A causes rapid loss of 47; **section on 74**; food chart 77; one apple equivalent to 1500 mg supplement 76; protects against high cholesterol 24; smoking hinders absorbtion of 25

Vitamin D section on **78**; food chart 80

Vitamin E excess iron hinders absorbtion 109; ineffective as supplement 143; **section on 81**; food chart 83; 62% reduction in heart disease when taken in natural form 143

Vitamin H See **Biotin**

Vitamin K section on **88**

Vomiting lack of Sodium 129

Walking excellent for improving health and mental outlook 116; cuts risk of cancer 20% 80; helps avoid headaches 87; important to prevent constipation 38; linked with longevity 122

Water plenty required for sustained energy 21; can leach Copper and Cadmium from pipes 25; lack of reduces muscle efficiency 41; soft water mildly acid 25

Water retention See **Edema**

Weak red blood cells See **Anaemia**

Weight average for NZ male/female 11; ruthlessly reduce to normalise high blood pressure 35

Weight control how to control intake of calories 153; fluid retention can be cause of weight problem 151; four essentials of 151; becoming permanently slim **section on 151**; rapid weight loss method 3 kg week 154

Weight loss lack of Thiamine 50

Williams Dr Ulric story of conversion to natural healing 141; criticism of modern medical system 146; healing advice from 149

Wind digestive enzymes lacking 72; excess Vitamin C 76

Words can affect our sense of well being 73; are they kind true and necessary 73

Wound healing slow lack of Vitamin C 75; lack of Zinc 135

Yeast Infections common following antibiotics 87; excess sugar in diet 45; lack of Iron 109; lack of Zinc 136

Yoghurt helps body re-establish necessary bacteria following antibiotics 87; more easily digested than milk 93; recipe for homemade 87; as virtually identical to milk not incl in food tables 13

Zinc alcohol can hinder absorbtion 15; double intake during pregnancy 11; Inisitol increases absorbtion 69; **section on 134**; food chart 137; lack of can raise cholesterol 25; excess can cause Copper deficiency 100; protects against excess Iron 110; protects against high cholesterol 24; raw bran can hinder absorbtion of 37; smoking hinders absorbtion of 25; Vitamin B6 necessary for absorbtion 58

Bibliography
(Three most recent updates only, although many other sources too numerous to list have been also utilised.)

NZ Food NZ People. Results of 1997 National Nutrition Survey.
NZ Ministry of Health 1999.

Recommended Dietary Intakes.
National Health and Research Council of Australia 1991.

Biochemical Effects of Dietary Boron in Animal Nutrition.
Curtiss D. Hunt, US Dept of Agriculture.

Doctor's Vitamin and Mineral Encyclopedia
Dr S.S. Hendler. MD. Aaron, 1991.

Healing Through Nutrition
Dr Melvyn R. Werbach MD. Thorsons, 1995.

Nutritional Medicine
Dr Stephen Davies & Dr Alan Stewart, Pan.

Good Cholesterol Bad Cholesterol
Dr Eli M. Roth MD, FACC. & Sandra L. Treicher RN, Compass Publishing, Australia.

The Vitamin Bible
Earl Mindell, Ph.B.,R.Ph, Arlington Books.

The Miracle Nutrient CoQ10
Dr E.G. Bliznakov, MD, & G.L. Hunt, Bantam, 1987.

Mental and Elemental Nutrients
Dr Carl Pfeiffer, Kents Publishing, Conn. USA.

Fit For Life
Harvey & Marilyn Diamond, Harper Collins, 1985.

Additive Code Breaker
Maurice Hanssen & Betty Norris, Lothian, 1992.

The Concise NZ Food Composition Tables
NZ Institute for Crop & Food Research, Dept of Health, Wellington.

Metric Tables of Composition of Australian Foods
Commonwealth Dept of Health, Nutrition Section, Canberra.

NZ's Greatest Doctor, Ulric Williams who became a Naturopath
Brenda Sampson. Privately published 1998. Wellington. Ph (04) 386-2514.

NZ Public Health Report
Ministry of Health, 1999.

Your Health at Risk
Toni Jeffreys Ph.D. 1998.

Everything you need to know about Collidal Silver
Max Crarer. Privately published 2000. Wairoa. Ph (06) 838-8660